Price $9.95

SINCERE'S
LAWNMOWER
Service BOOK

by
WILLIAM EWERS

Research Assistance
by
H. W. BAYLOR

PUBLISHED BY
SINCERE PRESS
POST OFFICE BOX 10422
PHOENIX, ARIZONA 85016

INDEX

Printed in the United States of America

First Edition 1971

Additional publications by Sincere Press

Sincere's Sewing Machine Service Book, 3rd Edition, William Ewers
Sincere's History of the Sewing Machine, William Ewers & H. W. Baylor
Sincere's Zig Zag Sewing Machine Service Book, William Ewers
Sincere's Bicycle Service Book, William Ewers
Sincere's Mini-Bike Service Book, William Ewers & Irv Charles

Introduction

Sincere How-To-Do books are unique for many reasons but the foremost must be because we concentrate on articles used in everyday living, by every family in the United States and most of the World. Machines that do wear, break down, get out of adjustment and require service can now be serviced in your home, by yourself, without mechanical training.

We explain and illustrate in basic, simple language how to service and adjust most malfunctions you will face. Knowledge of mechanics is helpful, but certainly not a necessity since most diagrams and illustrations are supplied by the manufacturers. When a clear diagram is not available, we obtain the part, or sub-assembly, and engage a commercial artist, to insure the utmost accuracy.

The research department always seeks full cooperation from important companies in the field covered. Most cooperate, some don't, but we never publish a book until a good cross-section of the industry is represented.

Our style and text may seem elementary to many of our readers, and it is meant to be. The text is simple because we know the book will be read by many people who know absolutely nothing about a lawn mower, or what makes it operate.

The young just-married home owner, the city dweller newly moved to the suburbs, the immigrant from another land, and the teen-ager embarking on his first business venture probably have never had the occasion to use a mower or understand its function. Over twenty years of working with machinery has taught me there is nothing too simple or basic to explain. Most serious repair jobs are the result of a minor problem left unattended and could have been avoided with a simple adjustment.

The age of ignorance about things mechanical is over. Mister average American now tackles the service job his counterpart of twenty years ago would never have attempted.

Why? What are the reasons behind the change? Technical training in our schools, from junior high school through college, is one answer. Advanced technology by manufacturers who turn out better built, simpler-to-operate machinery is another. Do-it-yourself projects are the "In"thing. And many tradesmen are pricing themselves out of business, or they can't keep pace with the growing population as fewer men enter the service business.

There are several points to consider before tackling your service job. First and most important, read the factory warranty thoroughly. Will the adjustment you're considering violate the manufacturers terms? If so, check with your local authorized dealer before proceeding. Most warranties cover defects in parts and workmanship for terms from 90 days to two or more years and simply mean that if you're negligent and break something, you are liable for service charges regardless of how old the machine may be. Follow engine warranty requirements to the letter. Any modification or change will be noticed immediately by the authorized agent and liability will be yours. After the warranty expires, it's a different story.

Use common sense and proper tools. Don't work on the equipment unless you can concentrate on the problem, and never work on a machine if you're angry or tired. There will be references to "special tools" throughout the book. Every tool mentioned may be purchased from an automotive jobber, and many are available on a one-time basis from the local equipment rental agency.

Finally, read the operating manual for your machine from cover to cover and follow all instructions. There are no short cuts.

William Ewers

Chapter One

The Reel Lawn Mower-Hand Operated

The important areas of service on the hand operated reel lawn mower are the reel/bed knife, and the wheel/pinion gear assemblies.

A dull blade or reel is more readily noticable on the hand operated reel mower because the operator can feel the difference while pushing the unit. Service should be frequent, especially on the lower-price models. Proper oiling and maintenance care will add years to the life of even the inexpensive machines.

Sharpening the Reel and Bed Knife.

There are two basic methods of sharpening the reel/bed knife assembly on the hand operated mower.

1. The basic makeup of the wheel/pinion gear assembly allows mower reel to turn when being pushed forward, but when pulled backwards, the reel is in neutral. Many of the expensive mowers utilize a shift-lever arrangement. When the lever is engaged, the mower reel will also turn in reverse.
2. To sharpen the reel/ bed knife on a mower equipped with the lever, loosen end bolts holding the bed knife and move it toward the reel. The reel should touch bed knife equally along entire surface of knife. Set the knife in place and tighten the end bolts.
    a. Most mowers are equipped with 2 bolts at each end.
    b. The 2 bolts allow for angle change to bed knife.
3. Spread either a valve grinding compound, or carborundum mixed with oil, over entire edge of bed knife.
4. Engage lever and pull the mower in reverse. Repeat several times until both reel and bed knife are sharpened.
5. Clean away grinding compound. Wash in gasoline or cleaning solvent until all compound is removed.
6. Position bed knife and turn bolts down fingertight.
7. Cut several strips of newspaper and place at each end and in middle of bed knife. Turn reel in direction of cutting. When properly set, the reel should slice the paper, without touching the bed knife.
8. Tighten end bolts.
9. If your mower isn't equipped with shift lever, follow the same steps and turn the reel manually. Use extreme caution and don't catch fingers between bed knife and reel. This process is slower and more care must be exercised to avoid serious injury.
10. Hand filing is the other alternative. The process is much slower,

and not nearly as accurate. Remove the bed knife and place in a vise. File along cutting edge and work slowly to attain an even edge.

11. Repeat on the reel, one blade at a time, without removing from the lawn mower.

Figure 1

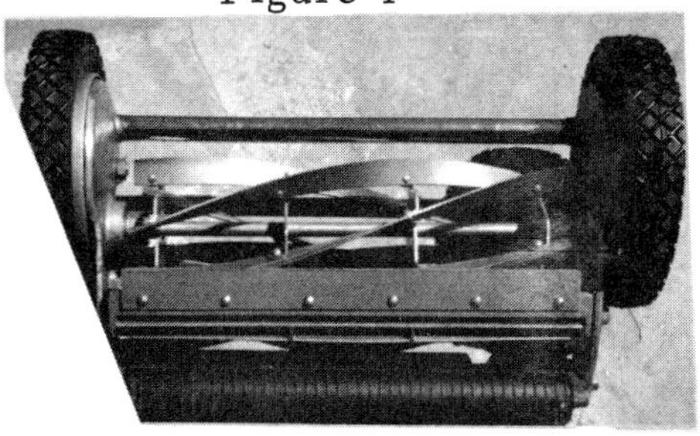

Figure 2

Adjusting Cutting Height.

1. To adjust cutting height on most hand mowers, change setting of the roller at back of unit. In most cases, there are 3 settings(figure 4).

Wheel-Pinion Gear Assembly.

1. To clean, lubricate, or replace, remove wheel hub, then loosen the axle nut (1), figure 3.
   - a. Remove the left wheel and pinion gear first. Grease with heavy viscosity grease. Clean all components before greasing. Grease the axle before replacing on mower.
   - b. Check the teeth in both wheel and pinion gear. If they are worn, broken or damaged, replace as needed.
   - c. Remove the right wheel and pinion, and repeat step a, above.
   - d. Never remove both assemblies at the same time. Left pinion will not operate with right wheel, and vice versa.
   - e. On units with shift lever, use care when removing unit. Replace all components in exact order of removal.

Lubrication.

1. Most points of lubrication are obvious. Oil holes denote points to oil but most mower owners don't oil often enough. Moisture from grass is a deadly enemy of the mower. Oil often, every time out is not too much.
2. Grease the wheel and pinion gear assembly at least twice per season.

Figure 3

Figure 4

Chapter Two

The Gasoline Engine.

Most power lawn mowers utilize the gasoline engine although the electric unit seems to be gaining in popularity.

This chapter covers basic fundimentals of two and four cycle engines, with emphasis on sub-assemblies such as carburetors, spark plugs, and a brief run-down on the magneto with breaker points.

Most manufacturers utilize one of the engines covered on the next several pages which includes Clinton, Tecumseh and Briggs & Stratton. A few companies manufacture the engine as well as the mower and we've covered them on a relative basis within the chapter.

Since the majority of service problems on the power mower are engine oriented, it would be a good idea to study this chapter in depth.

Additionally, there is some excellent material covering a complete and thorough overhaul in a later chapter.

Theory.

The internal combustion engine is a reliable source of power. When it is combined with the lawn mower, the combination is nearly perfection.

Both two cycle and four cycle engine theory is explained in easy-to-understand terms.

The Briggs & Stratton Engine.

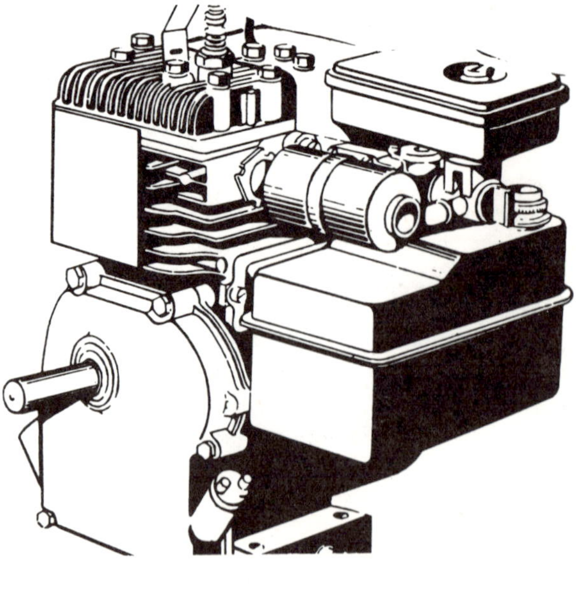

The running engine creates heat. But how? The old saying, "Where there's heat, there's fire," must also be true in the gasoline engine.

One simple way of explanation would be an illustration between a cannon and an engine. Everyone knows how a cannon operates.

1. Light the fuse. Spark plug on the engine.
2. The powder burns. Gasoline and air mixture in the engine.
3. Expanding gasses force the cannon shell out the gun barrel. Piston in cylinder is forced down in the gasoline engine.

Figure 5

The spark plug.

The concept is basically simple. The shell leaves the cannon barrel on it's way toward an objective. It's entire energy is expended in one blow. One blow won't move the mini-bike along. Additional components are required to keep shell(piston) from leaving the barrel(cylinder).

First requirement is a connector to keep piston from flying out the end. A connecting rod operates much like your legs when you're riding a bike. One end is bolted to the piston, the other is held in place on the crankshaft by a crank pin.

The piston is connected to the crank shaft so it can return to a starting position for another shot of gasoline and air mixture, which when fired by another spark from the spark plug, will drive it down the cylinder again, producing the power to move the crankshaft and whatever it powers.

Figure 6

Piston, Connecting rod, and Crankshaft. Steen's

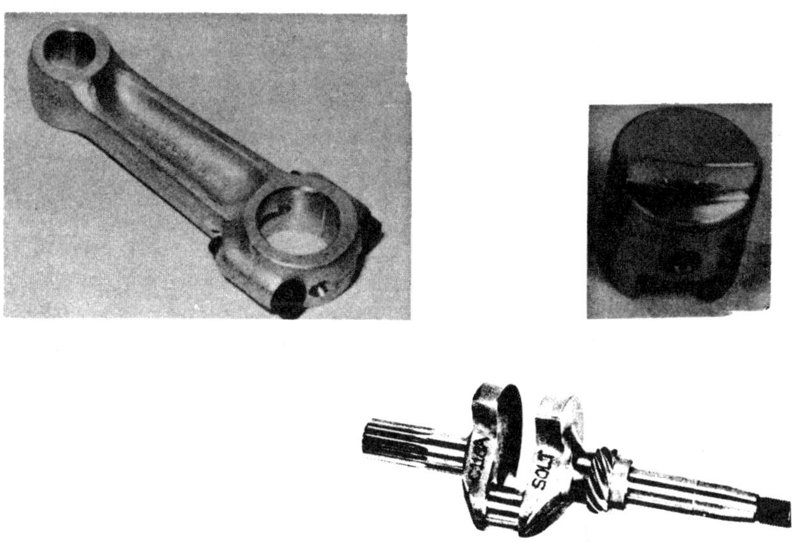

Figure 8

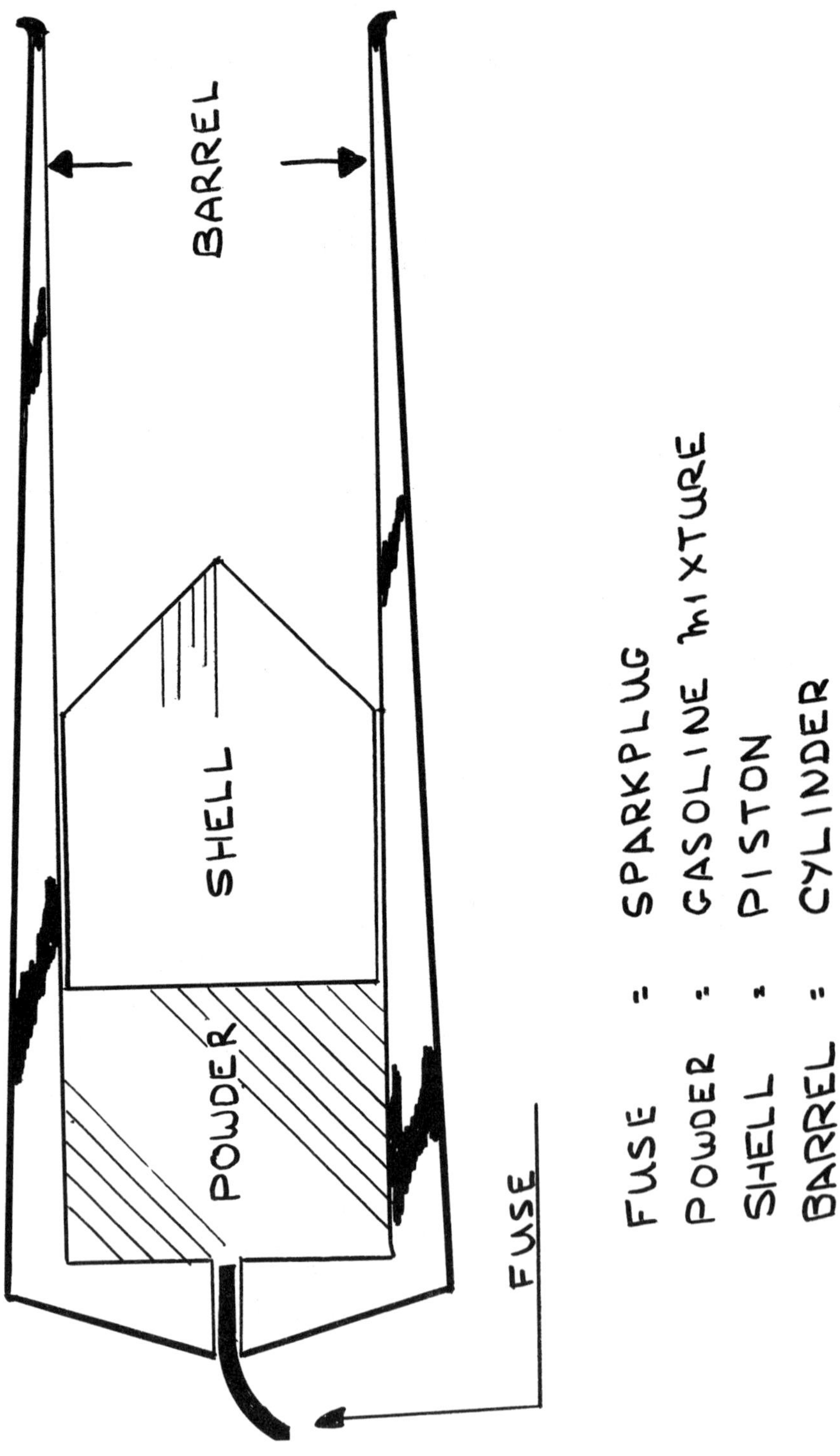
BARREL
SHELL
POWDER
FUSE
FUSE = SPARKPLUG
POWDER = GASOLINE MIXTURE
SHELL = PISTON
BARREL = CYLINDER

Figure 9

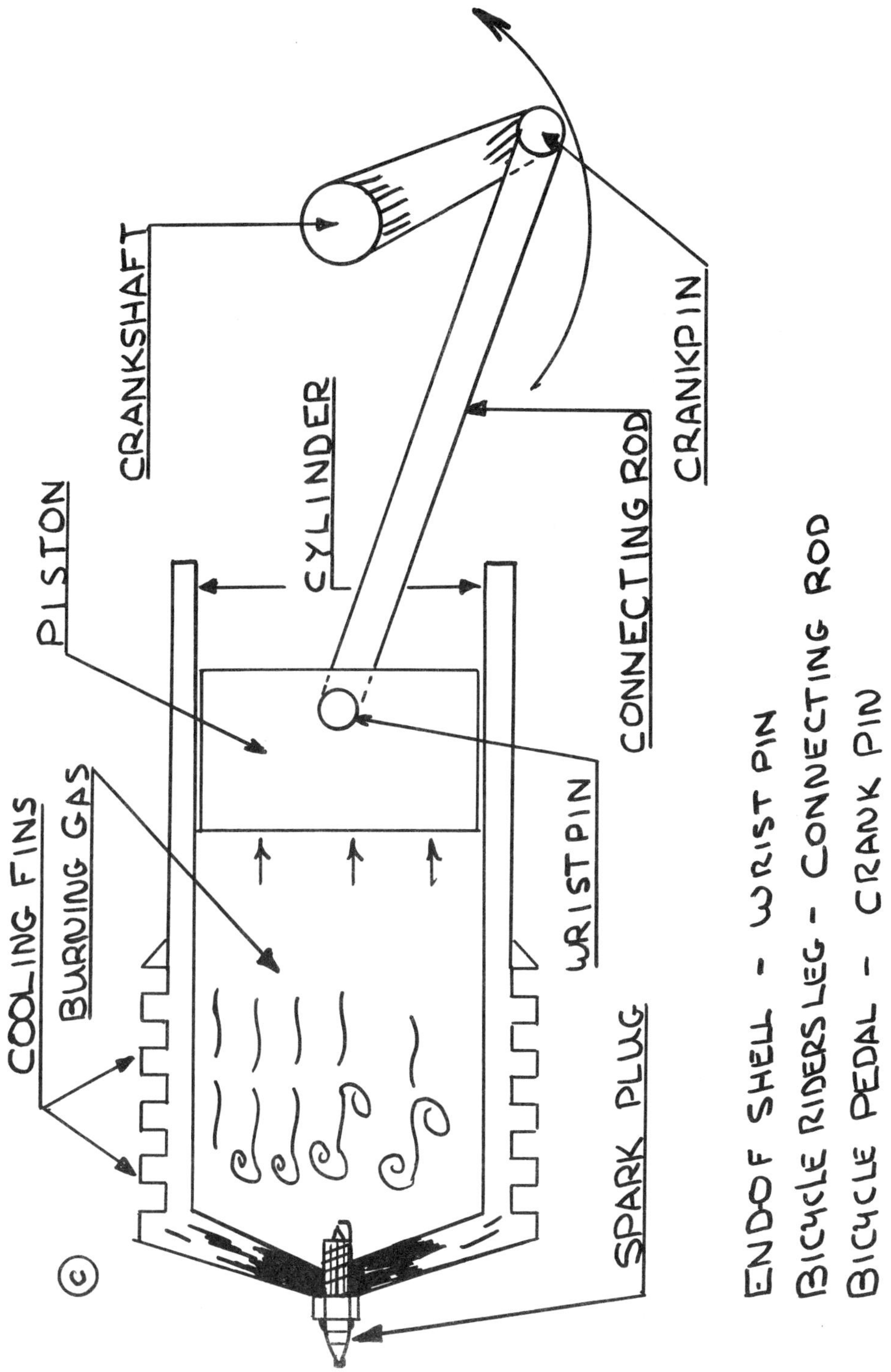
COOLING FINS
BURNING GAS
PISTON
CRANKSHAFT
CYLINDER
CONNECTING ROD
CRANKPIN
WRISTPIN
SPARK PLUG
©
END OF SHELL - WRIST PIN
BICYCLE RIDERS LEG - CONNECTING ROD
BICYCLE PEDAL - CRANK PIN

For further comparison we'll borrow again from the principal of cycling.

Center shaft of a bike sprocket	Crankshaft of an engine.
Sprocket from a bicycle	Drive gear of an engine.
Chain	Drive chain.

Figure 9

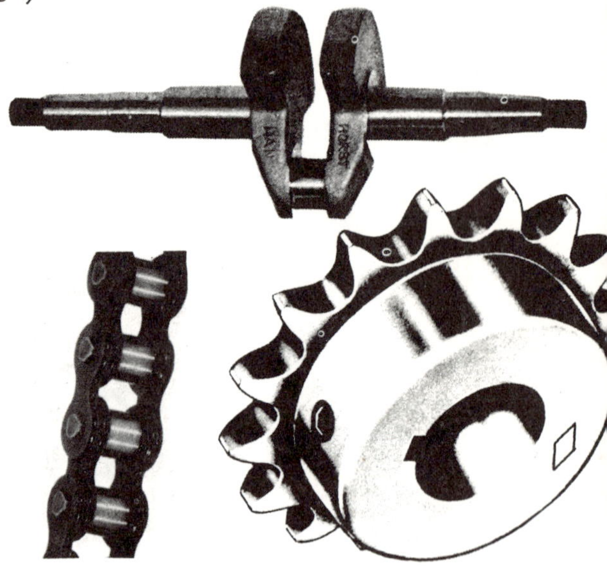

To move along, it requires more than one pump of the pedal, or surge of power from the engine. There must be a way to get a new mixture of gasoline and air in, and the burned exhaust gasses out.

Two holes near the top of the cylinder are called ports. If there's no way to close the ports, all the pressure will escape when the mixture of gasoline and air burn, instead of pushing the piston and producing power.

A plug is utilized for this job. It's called a valve, since it opens & closes as needed. The valve head is attached to a rod, with a spring to supply needed pressure to keep the valve closed as required. The rod(valve stem) is activated by the cam.

Figure 10

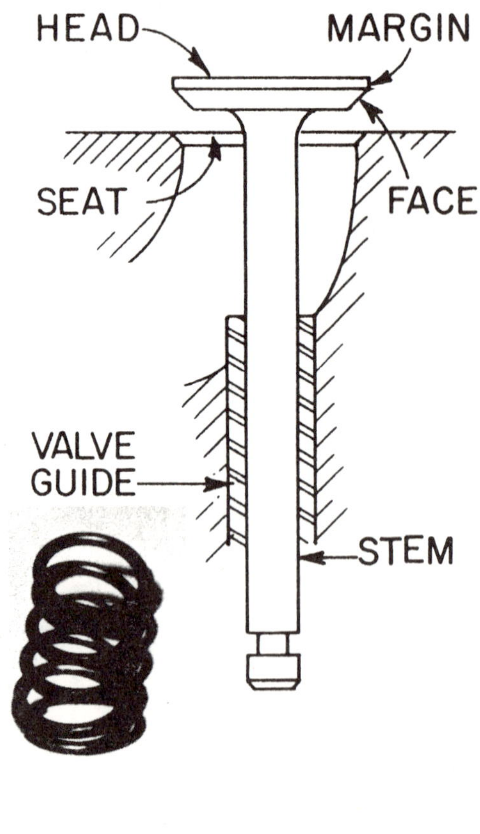

How the Gasoline Engine functions:

There are two different type engines. The two stroke is pictured below and the four stroke is pictured on the following page.

They look very similiar. In fact, many parts interchange on some units. The two stroke has no valves and requires a mixture of gasoline and oil to operate. It burns fuel on every stroke and produces power on every stroke. It has only three basic moving parts (crankshaft, piston and rod) which means less service, and more efficiency.

The four stroke (or cycle) performs a different function on each of the four strokes. First is the intake, then the compression, then the power, and finally, the exhaust stroke. Valves are utilized, and the engine burns regular gasoline. Lubrication is from a different source in crankcase.

Figure 11

Two stroke Clinton Engine

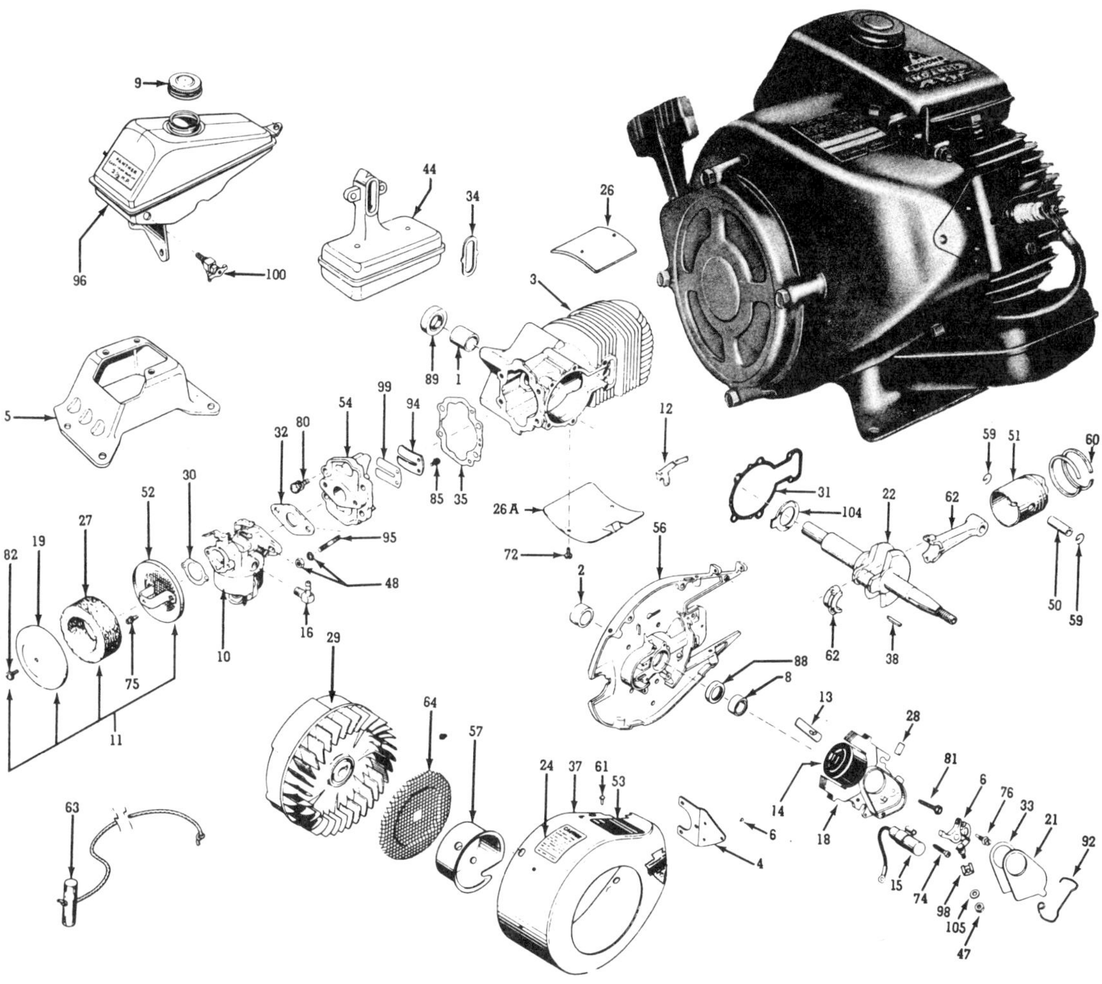

Figure 12

Four cycle Briggs & Stratton Engine

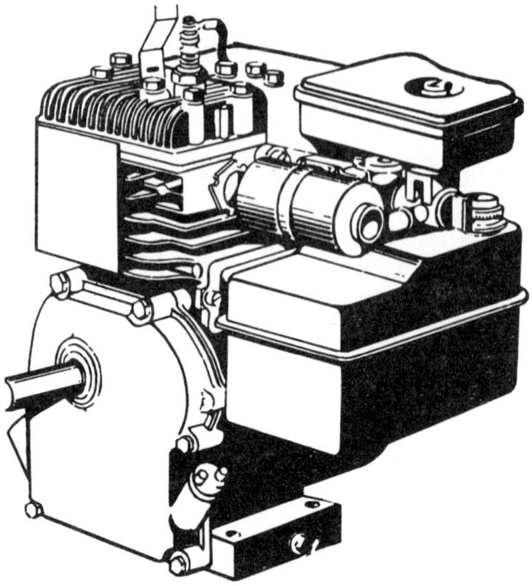

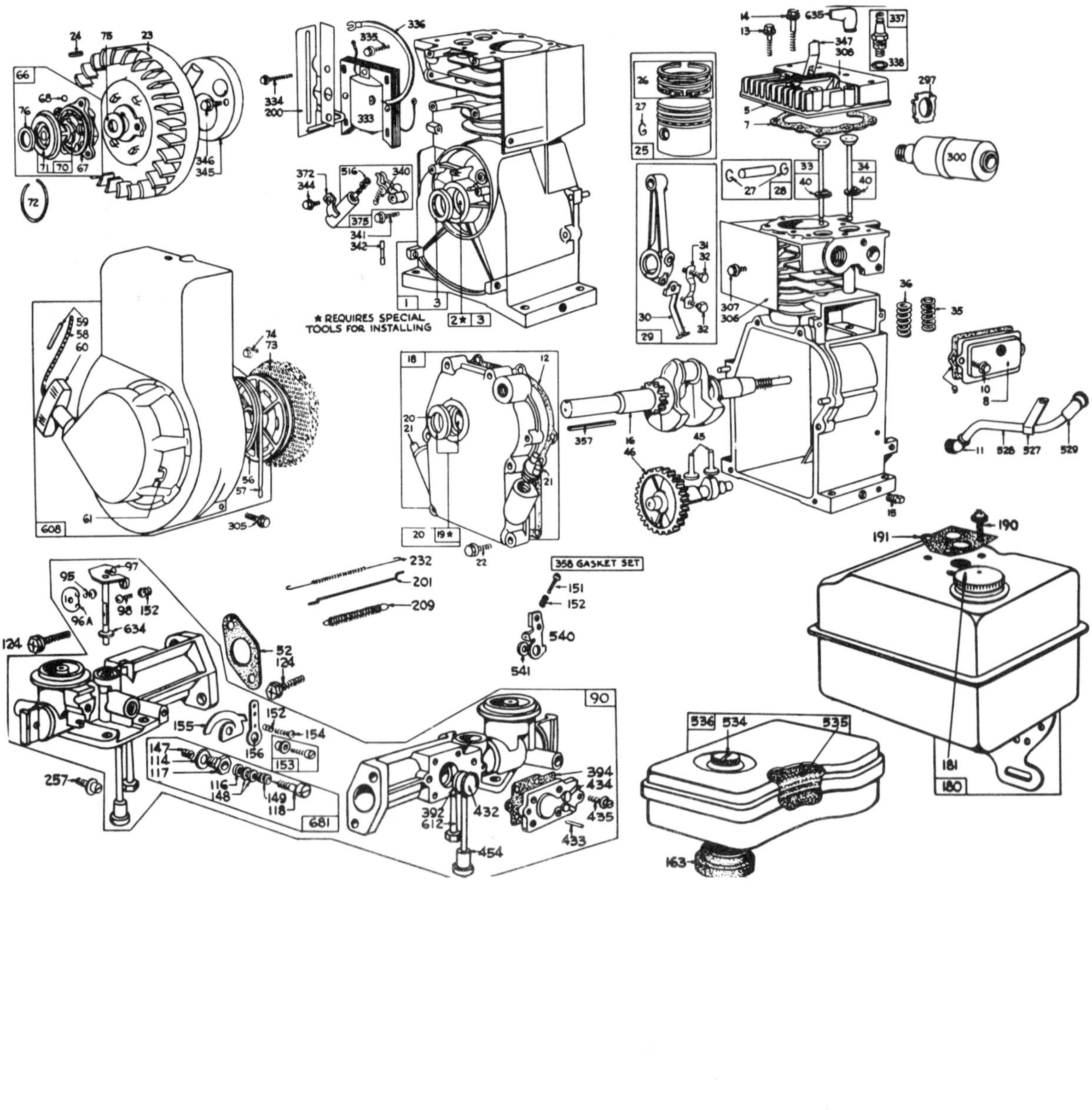

Now, back to the valves and their function. There are two, intake and exhaust, on every four stroke engine.
The intake valve opens to allow gasoline and air mixture in.
The exhaust valve opens to allow burned gasses out.

The Four Stroke Engine:
Intake stroke;
The piston is at the top of cylinder (top dead center). As it starts down the intake valve opens, allowing gasoline and air mixture in. Lowering of the piston creates a vacuum in the cylinder and outside air pressure forces the mixture into the cylinder.

Figure 13

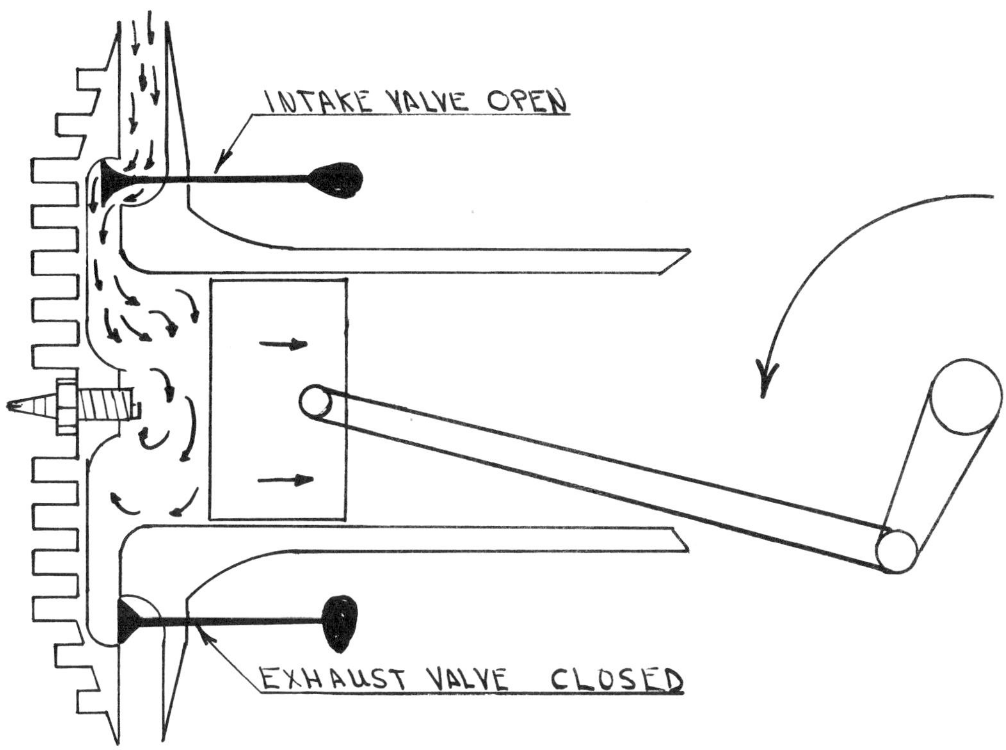

INTAKE STROKE

Compression stroke;

The piston is at extreme bottom (bottom dead center). As the piston rises the intake valve closes. The piston presses the gasoline and air mixture together. When the piston arrives back at the top of the cylinder, the gas is fully compressed. The timing mechanism activates the ignition system which in turn causes the spark plug to spark, igniting the gas in the cylinder.

Figure 14

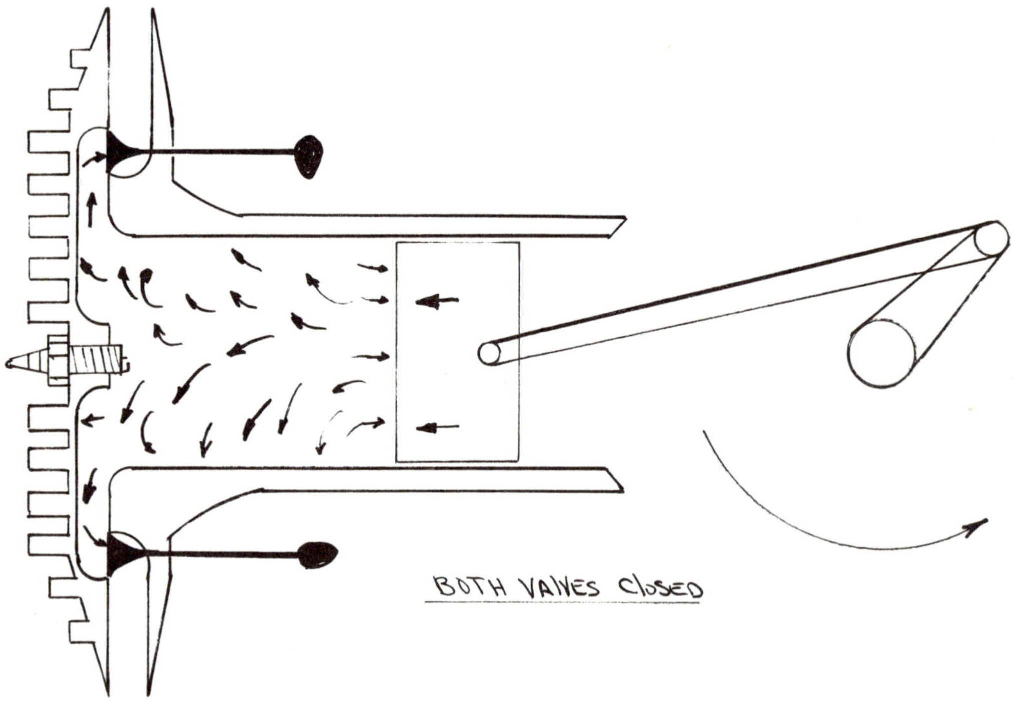

COMPRESSION

Power stroke:

The ignition occurs when the piston hits top dead center, lighting off the burned gasoline and hurling the piston downward with a blast, delivering power to the crankshaft, which in turn delivers the power to the drive shaft.

Figure 15

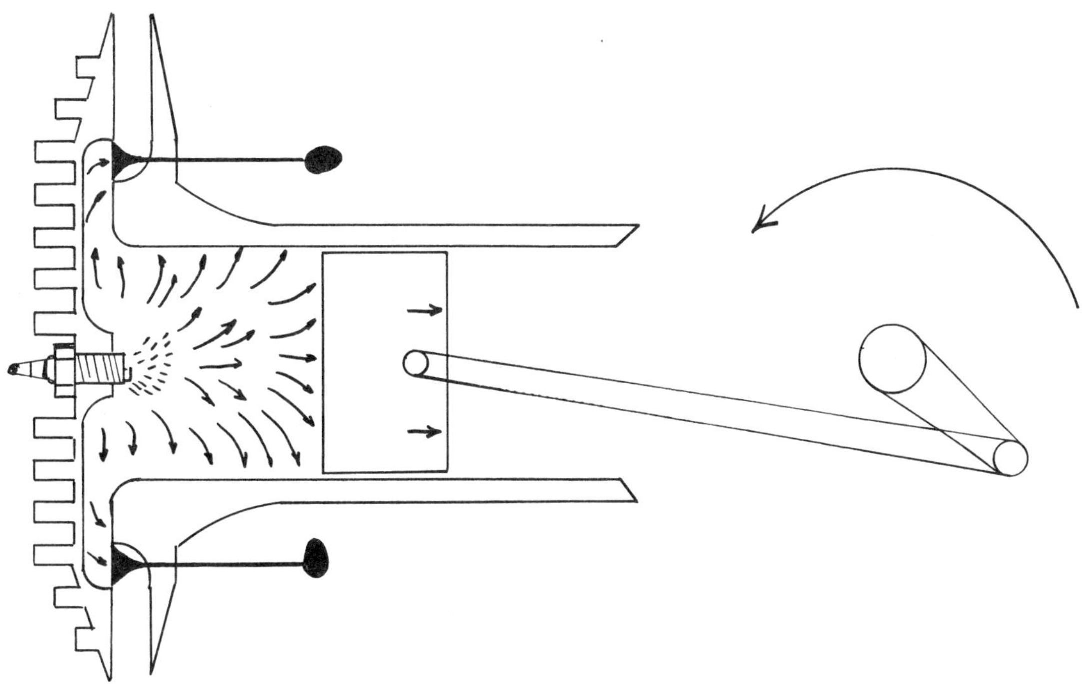

POWER STROKE

Exhaust stroke:

The piston is now at bottom dead center again and the engine must dispose of all the burned up gasoline. As the piston rises, the exhaust valve opens and the burned gasses are forced out. When the piston reaches the top, the exhaust valve closes. When the piston starts back down, the intake valve opens again, and the cycle is repeated.

Figure 16

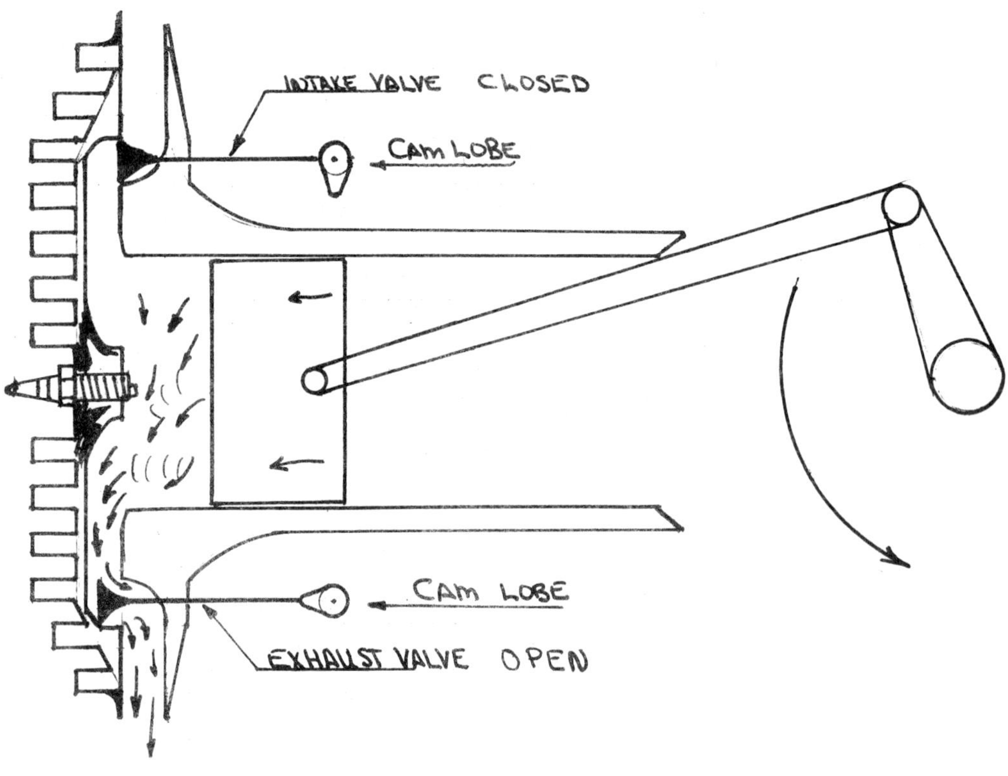

EXHAUST STROKE

The Two Stroke Engine:

The two stroke engine uses a combination of oil and gasoline for fuel. Mixture is important and manufacturers suggestions must be followed.

In the two stroke engine, intake, compression, power and exhaust are completed in two strokes of the piston.

As the piston moves upward on it's stroke, it passes the intake port, closing the port opening. Continued upward movement causes the gasoline and oil mixture, together with air, to be compressed. At the same time a new fuel charge is drawn into the crankcase. When the piston nears the top of the compression stroke, the fuel mixture in the chamber is ignited by the spark. The explosion, and expansion of gasses, force the piston down on the power stroke.

As the piston nears the bottom of the stroke, it uncovers the exhaust port opening, slightly ahead of the intake port, which permits taking advantage of exhaust gas pressure in the chamber, allowing them to start escaping. Continued downward travel of the piston uncovers the intake by-pass port. The incoming charge assists in forcing exhaust gasses out of the cylinder to complete the cycle.

Figure 17

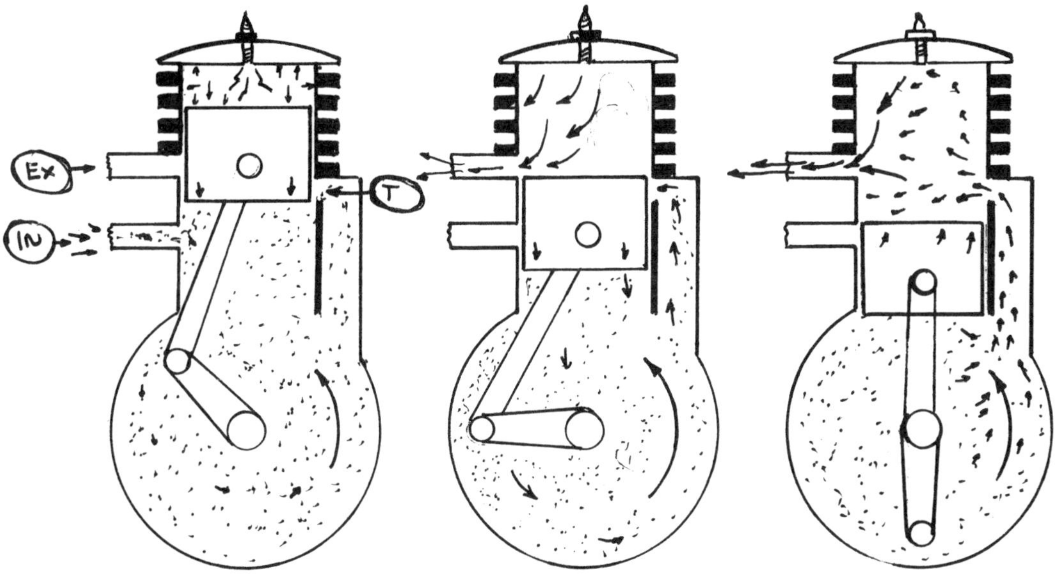

TWO STROKE CYCLE

EX- EXHAUST PORT IN- INTAKE PORT

T- TRANSFER PORT

Carburetion:

The proper amount of air and gasoline, mixed together to produce gasses and power the engine, is called carburetion. Gasoline comes from a tank on the engine, but where is the source for air? And how are the gasoline and air mixed in correct proportions?

Figure 18

Carburetor

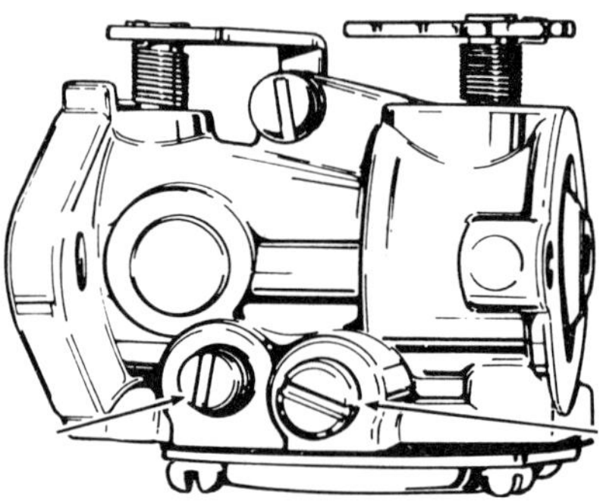

Have you ever blown across the open top of a full glass of water? What happened? Some of the water was blown out. We have a similiar situation in carburetion. It's necessary to find some method of keeping the blowing constant, and still keep the glass full.

First, narrow the width of the glass at the mouth and rename it a "jet." Next, place a tube in the opening and call it the carburetor throat. Now, hook it to a can of gasoline by means of a tube from the bottom to the bottom of the jet. See figure 19.

Figure 19

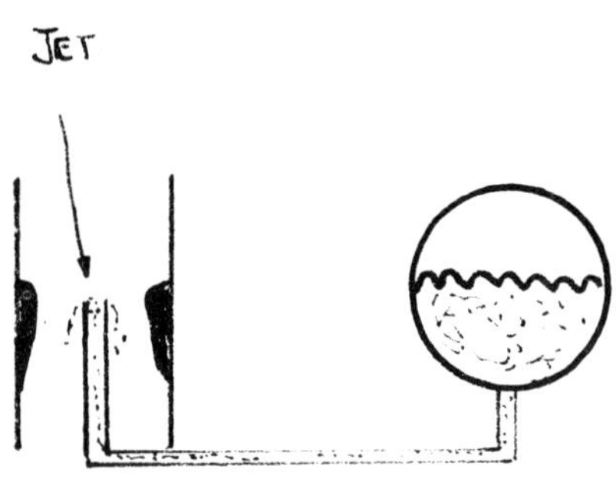

GASOLINE FEED

After the two are connected, it's necessary to control the gasoline flow so it doesn't pour through the open jet. We just want to keep it level full.

Place a float inside the can and set it so it stops the flow of gasoline at a given level.

When this part is connected to the intake port, the carburetor will run wide open unless some control is attached. Close the tube and it will draw less air, and slow down the operation. Add a flat disc inside, and fit it so it closes the hole, and you have a control called the throttle valve.

To review, the engine won't operate unless a combustable fuel charge is supplied to the cylinder at precisely the right time. In a carburetor-type engine, this fuel charge is formed by a mixture of air and vapor resulting from a vaporization of gasoline. The process of vaporizing gasoline, and mixing it with the proper amount of air, is called carburetion.

Figure 20

Venturi action.

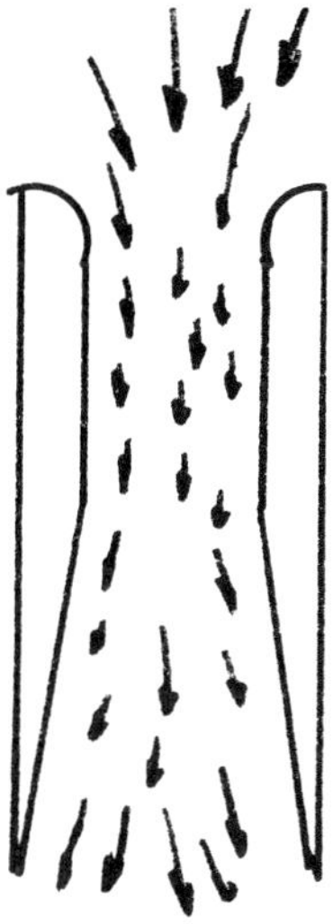

To understand why air moves from one area to another, it's necessary to understand something about pressure.

Try it yourself. Take a mailing tube, which is cylindrical, and try this simple test.

With both ends open, the atmospheric pressure at all areas around and inside the tube are equal. Therefore, no air movement.

Insert a solid cylinder inside the tube, then pull it out again. As the solid cylinder is removed you'll notice air movement into the open end of the tube. The air flow is caused by low atmospheric pressure behind the solid cylinder which causes the outside atmospheric pressure to rush into the open end of the tube to balance out the low pressure.

To illustrate further, place a piece of paper over the end of the mailing tube and remove the solid cylinder again. You'll notice that outside pressure will force the paper inside the tube. Use lightweight paper to test.

In carburetion the artificially created low atmospheric pressure caused by downward travel of the piston in the cylinder, is used to facilitate movement of air and fuel into the combustion chamber.

Figure 21

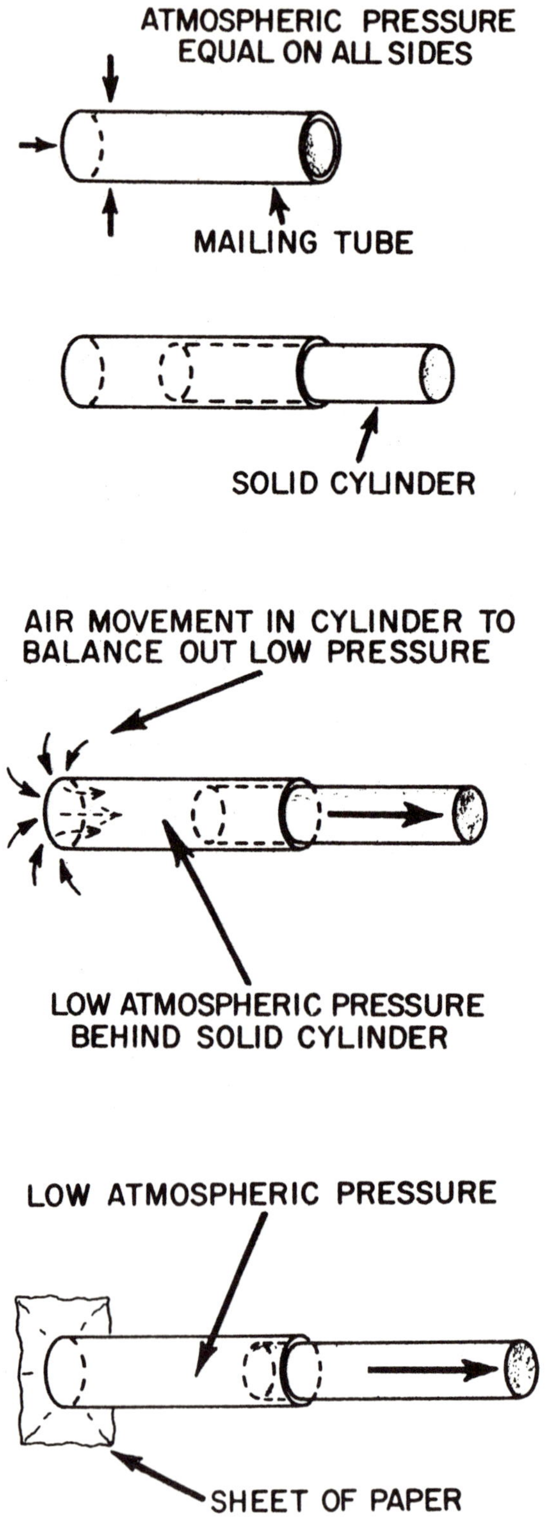
ATMOSPHERIC PRESSURE
EQUAL ON ALL SIDES
MAILING TUBE
SOLID CYLINDER
AIR MOVEMENT IN CYLINDER TO
BALANCE OUT LOW PRESSURE
LOW ATMOSPHERIC PRESSURE
BEHIND SOLID CYLINDER
LOW ATMOSPHERIC PRESSURE
SHEET OF PAPER

Figure 22

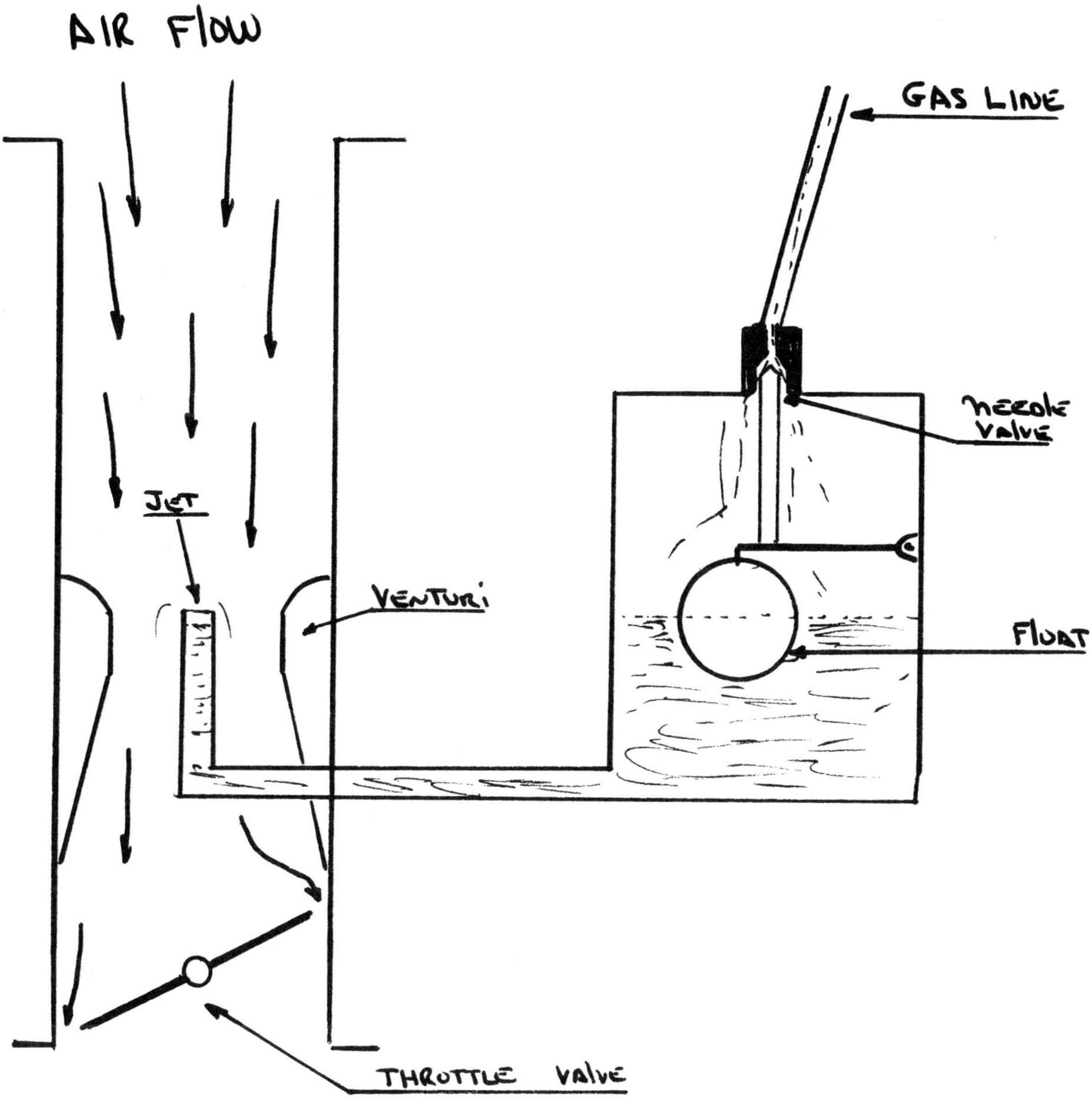
AIR FLOW
GAS LINE
NEEDLE VALVE
JET
VENTURI
FLOAT
THROTTLE VALVE

Proper carburetion is essential to good engine operation. Most engines have a high speed adjusting needle in addition to the idle adjusting needle. Adjustment is necessary when peak performance isn't attained when the throttle is opened.

The maximum setting for most small engines is 4000 RPM, and is more than adequate for good performance. An attempt to set it higher might result in permanent motor damage.

Float type Carburetor.

The high speed circuit is activated when the engine is operating at working speeds, and is important to good carburetor operation because it must supply a combustable mixture to the engine under full, partial, or no load conditions, and at various engine speeds. The circuit consists of a metering jet, high speed adjustment screw, main nozzle, and air jet. Its principal is simple, starting with atmospheric pressure in the throat of carburetor, which is created by downward travel of the piston in the cylinder when the throttle butterfly is nearly, or completely wide open. With low pressure now concentrated behind the venturi, fuel is forced through the metering jet, around the high speed adjusting screw, into the main nozzle where air from the air jet blends with the fuel. From there the fuel-air mixture moves up into the throat, where additional air coming in from the front of the carburetor helps move the fuel-air mixture into the combustion chamber. The amount of fuel is controlled by the high speed adjustment screw. The carburetor throttle assembly controls engine speed by changing pressure in the main nozzle area, which when reduced, will slow down the movement of fuel-air mixture through the high speed circuit.

Figure 23

High speed circuit.

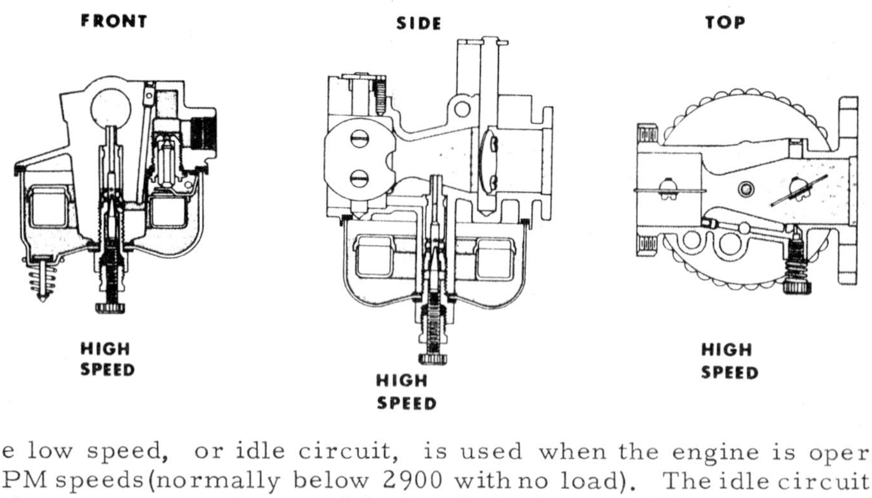

The low speed, or idle circuit, is used when the engine is operated at low RPM speeds(normally below 2900 with no load). The idle circuit is calibrated so a proper mixture of fuel and air is supplied at an RPM range of from 1400 through 2900. This circuit consists of a fuel metering jet, air jet, primary idle port, intermediate idle port, and the adjustment screw. The circuit starts operating when the throttle butterfly is closed, or nearly closed, closing the throat of the carburetor. At this point the low atmospheric pressure is concentrated behind the throttle butterfly, causing fuel to move through the fuel metering jet, into the idle passage, where air is mixed with fuel. The fuel is now ready to move into the primary, or intermediate, port area of the carburetor. The amount of fuel-air mixture is controlled by the idle adjustment screw, located in the primary port area.

The primary idle port operates apart from the intermediate, up to about 2200 RPM, no load, when the intermediate ports begin to supply the fuel mixture to the idle circuit. If the intermediate port fails to function, there will be a flat spot in engine operation between 2000 and 2900 RPM's when the high speed circuit cuts in.

Figure 24

Idle Circuit.

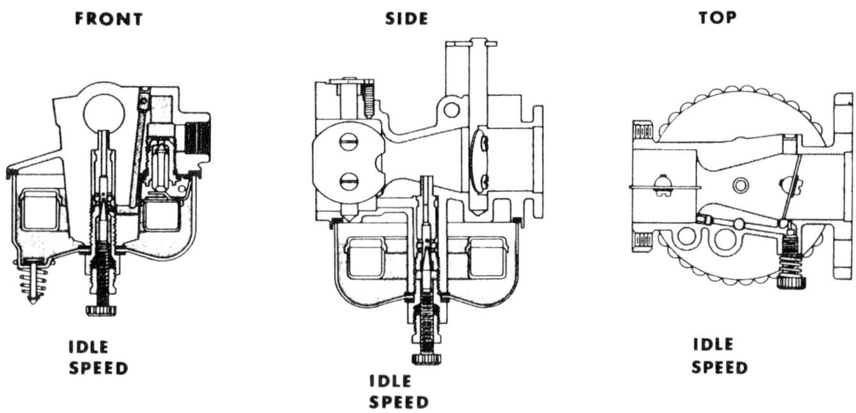

The float controls the level of fuel in the carburetor bowl. When it's properly adjusted, sufficient fuel is available for both the idle, and high speed circuit operation. It consists of a metal float, inlet needle, and the seat. The inlet needle and seat are calibrated to control level and amount of fuel supply to the bowl, with either a fuel pump, or gravity feed fuel supply. The float level setting is very important. Any maladjustment will put entire system out of balance, making the engine hard to start, prevent it from idling properly, and most important, cut down total performance.

A choke is necessary for starting the engine in different climatic conditions, and with different fuel mixtures. The choke is located in the air-intake side of the carburetor and consists of a shaft and butterfly. When activated, it can close or almost the throat of the carburetor and create a low pressure area in the venturi. As the butterfly closes, a reduced amount of air is supplied to a greater amount of fuel as it moves into the venturi, which forms a richer fuel mixture and allows engine to start easier.

Always avoid overchoking, or prolonged choking. The rich mixture will wash away oil film from cylinder walls, score the piston, and dilute crankcase oil.

Figure 25

Choke Circuit.

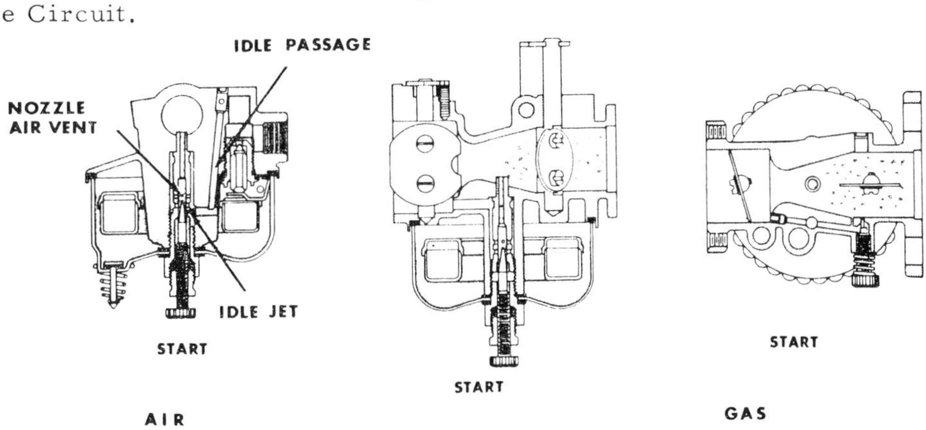

Carburetor adjustment- Tecumseh-type engine.

If you bought a new unit, the carburetor was adjusted at the factory, but may have changed in transit.

The carburetor high speed setting must be made with engine at full speed.

To adjust, proceed as follows:

1. Close high speed adjusting needle finger tight. Turn clockwise . Don't force if it doesn't turn easy.
2. Open it counterclockwise, one full turn.
3. Close idle adjusting needle finger tight. Turn clockwise.
4. Back it off about 5/8 of a turn, counterclockwise.
5. Start the engine.
6. Turn control to full throttle and adjust high speed needle about 1/8 turn in either direction until engine runs smoothly. Allow engine to run at each setting for at least 10 seconds, to give it time to react. If engine runs smoothly, setting is correct.
7. Close throttle until engine is idling and adjust idle needle in the same manner.

Figure 26

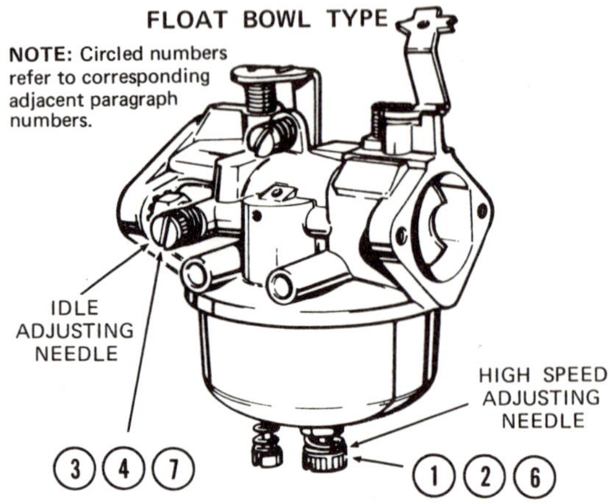

DIAPHRAGM TYPE

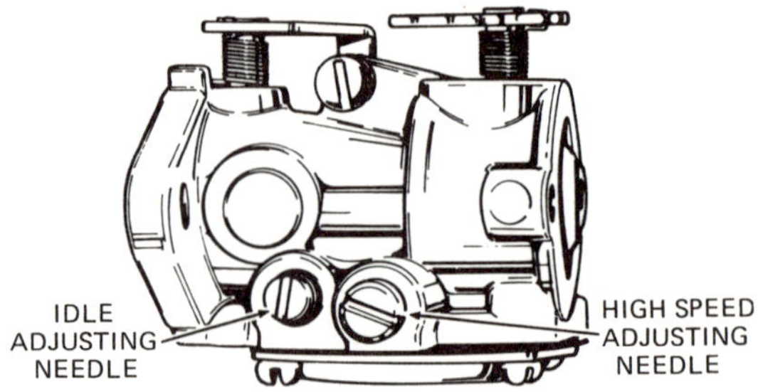

If the engine runs rough after you've had it for some time, check the air cleaner. A dirty or partially clogged air cleaner will cause engine to run rich and lose power.

Remove the air cleaner unit and tap against a solid surface. This will remove all foreign matter and dirt accumulation.

Never wash a paper air cleaner, or try brushing dirt away. Either of the practices will destroy the filtering ability.

Never run an engine without a filter in place. Never run an engine with a filter that's been punctured.

If there's any doubt about the filter condition, buy a new filter. It's less expensive than an engine overhaul.

Figure 27

Paper element.

There are three basic areas of malfunction in the carburetor, too lean, too rich, or fuel leakage. Efficiency can also be hindered by any foreign matter, liquid or solid, that interferes with the proper flow of air or fuel.

If the problem is serious, broken parts etc., take it to your dealer, especially if you have no mechanical experience. However, if you would like to do the job yourself, obtain a service manual from the maker of the engine, and give it a whirl. This book will help guide you.

The Carter carburetor is illustrated below, but basic service procedures will apply to most units.

Dis-assembly.

1. Remove carburetor from engine.
2. Remove bowl nut, gasket and bowl.
3. Remove float pin, float, needle and needle seat. Check float for dents, leaks, or wear on float lip or float pin holes.
4. Remove bowl ring gasket.
5. Remove low speed jet, and the high speed adjusting needle assembly, and the spring.
6. Remove idle adjustment screw and spring.
7. Remove nozzle.
8. Remove throttle valve screws, valve, shaft, and lever assembly.
9. Don't remove choke valve and shaft unless replacement is needed. A spring-loaded ball holds the choke in a wide open position. Replace both ball and spring when replacing choke shaft and lever unit. Hold a screwdriver or small piece of wood over threaded hole in air horn(opposite side from choke lever), to prevent the ball from coming out when shaft is removed.
10. Clean all parts in a cleaning solvent. Be sure all carbon is removed from bore, especially where throttle valve seats to the casting. Blow out other passages with compressed air if available. Replace all worn or damaged parts. Al- replace the gaskets.

Figure 28

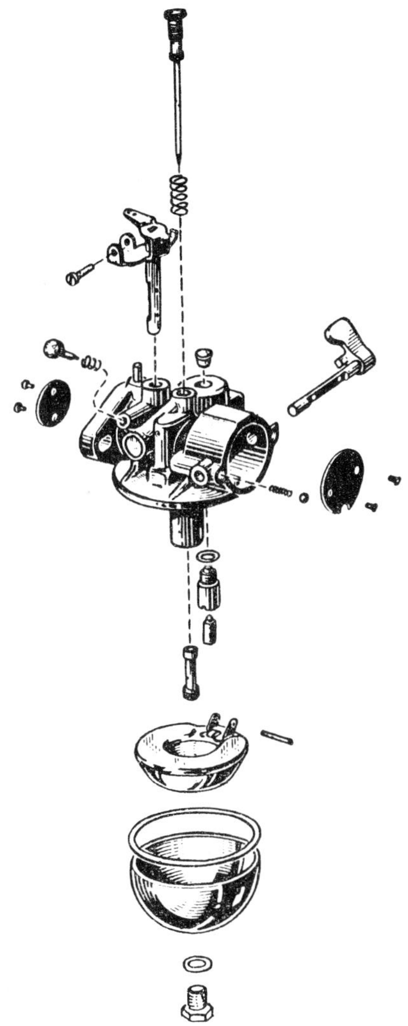

Re-assembly.

1. Install throttle shaft and valve. Valve must be installed with trademark "C" on side toward the idle port when looking at flange side. Always replace with new screws if possible.

2. With valve screws loose, and throttle lever set screw backed off, seat the valve by tapping lightly with a small screwdriver. Hold in place while tightening the screws.
3. Install nozzle and be certain it seats properly in the casting.
4. Install needle seat, needle, float and float pin.
5. Set float level. With carburetor casting inverted, float resting lightly against needle in its seat, there should be 3/16" clearance between machine surface of casting and free end of float (side opposite needle seat). Adjust by bending top of float with a small screwdriver.
6. Install bowl ring gasket, bowl, bowl nut gasket and bowl nut. Tighten securely after centering bowl in gasket.
7. Install low speed jet and high speed needle assemblies. Turn inward until it seats in nozzle, then back it out about two turns.
8. Install idle adjusting screw fingertight. Back out about 1 1/2 turns.

Figure 29

Tillotson HL series carburetor-parts breakdown and complete unit. Steen

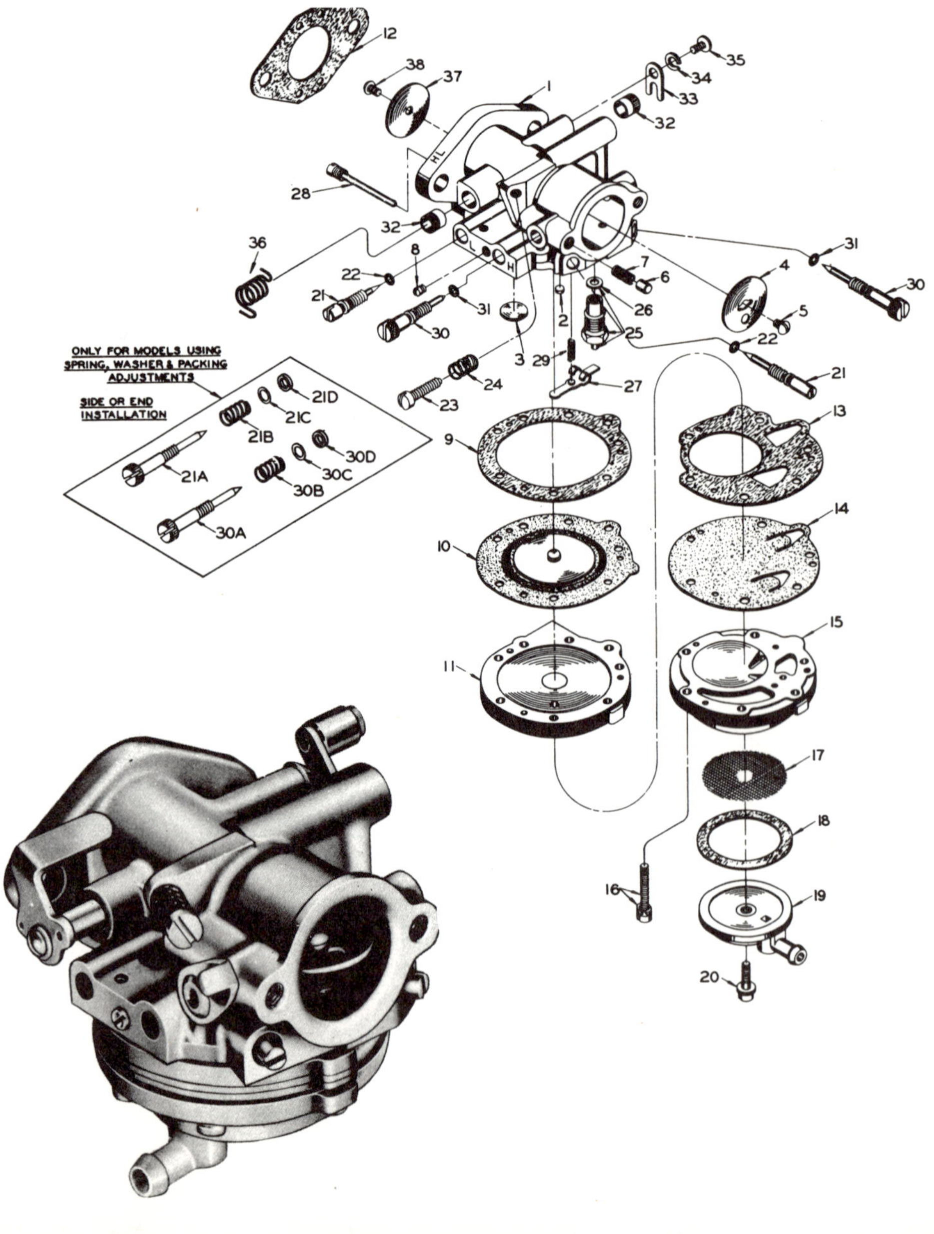

Carburetor, air filter and gas tank-Briggs and Stratton engine.

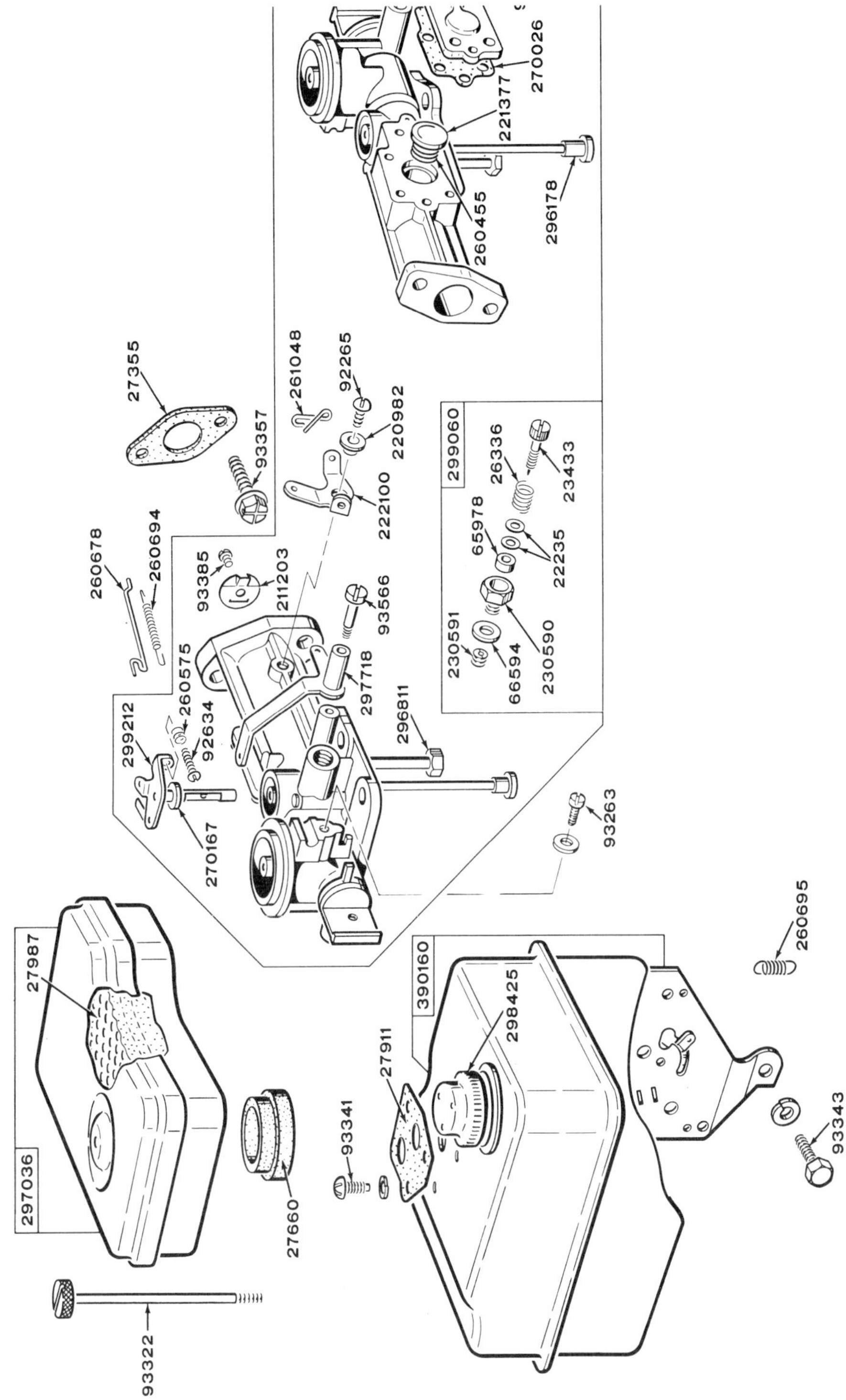

Figure 31

Parts breakdown- Briggs & Stratton

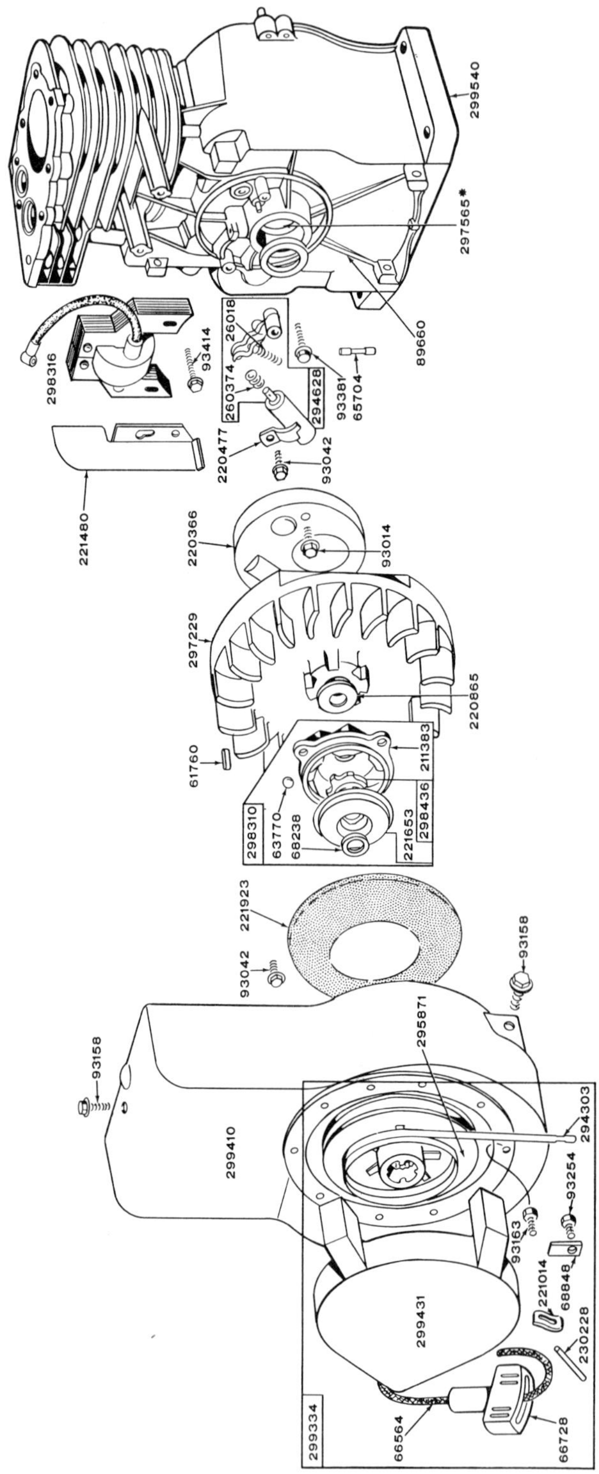

Figure 32

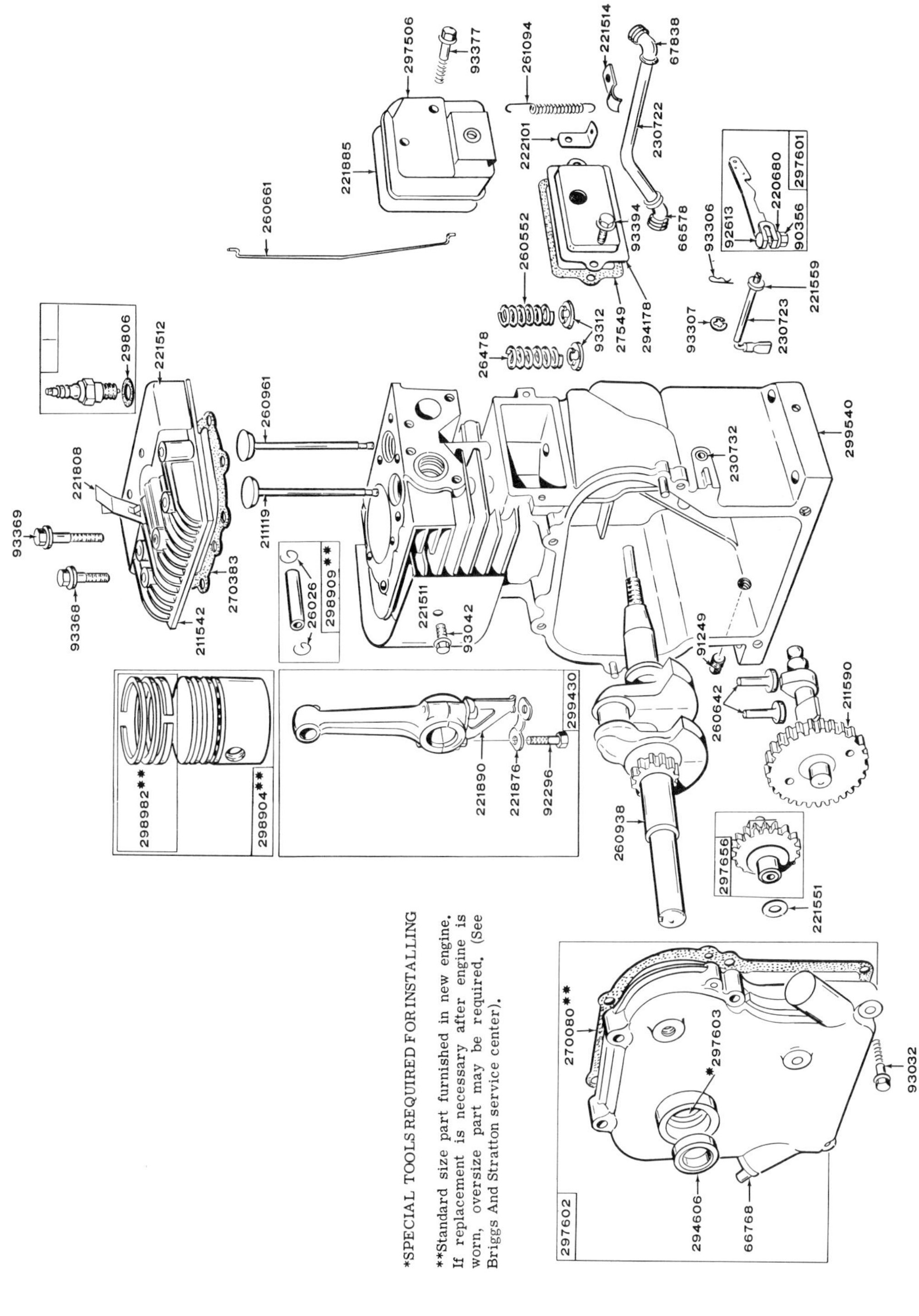

*SPECIAL TOOLS REQUIRED FOR INSTALLING

**Standard size part furnished in new engine. If replacement is necessary after engine is worn, oversize part may be required. (See Briggs And Stratton service center).

The Spark Plug:

A good spark plug is essential to trouble-free engine operation. If the spark doesn't get down to the fuel, it can't ignite and produce the needed function.

The spark plug should be checked frequently. Remove from the engine and visually inspect the electrodes. Proper gap is illustrated below, for each different type. On two cycle engines the gap should be from about .028 to .033. On four cycle engines the gap should be from about .025 to .028.

If the spark plug has heavy carbon deposits, burned electrodes, or has cracked insulation, replace it.

The plug is probably functioning normally if brown to greyish-tan deposits are on electrodes. If slight wear is present, re-gap and replace in the engine.

If the plug has eroded electrodes and a pitted insulator, replace with a new plug because old one is near end of useful life.

Carbon-fouled plugs are easy to spot. Dry, fluffy, black deposits are generally the result of over-rich carburetion and excessive choking. The clogged air cleaner can also restrict air flow to the carburetor, causing the rich mixture. Poor ignition output may also be the culprit. Faulty breaker points, weak coil or condenser, worn ignition cables, can also reduce voltage and cause misfiring. Excessive idling, slow speeds under light loads may keep plug temperatures too low so normal combustion deposits don't burn away. The problem can be cured with a hotter burning plug.

The oil-fouled plug is generally the result of worn rings which allow oil to be pumped by. Merely clean oil-fouled plugs and use them again.

Burned or blistered insulator nose and badly burned electrodes are indicative of spark plug overheating. Improper spark timing, or very low octane fuel can cause detonation and overheating. Cooling system obstructions or sticking valves may also cause this condition, A fuel mixture that is too lean may also be the problem.

Figure 33

Figure 34
Normal plug

Figure 34a
Worn plug

The Magneto.

The magneto is a stationary self-contained generating unit consisting of an armature plate assembly, condenser, and breaker point assembly.

If the spark plug has been replaced, and spark still appears weak, check magneto output. When troubleshooting the small gasoline engine, one of the first areas to check should be magneto output.

Clinton Engines Corporation suggest the test be made with a wide-gap spark plug, .156" to .187" is suggested.

1. Fasten the high tension wire lead to the plug, and ground plug on the engine head.
2. Crank the engine and observe. If the spark doesn't jump, additional magneto service is required.

Figure 35

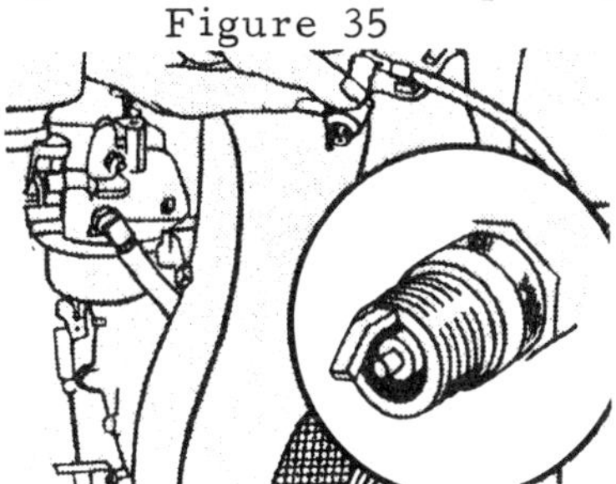

Figure 36

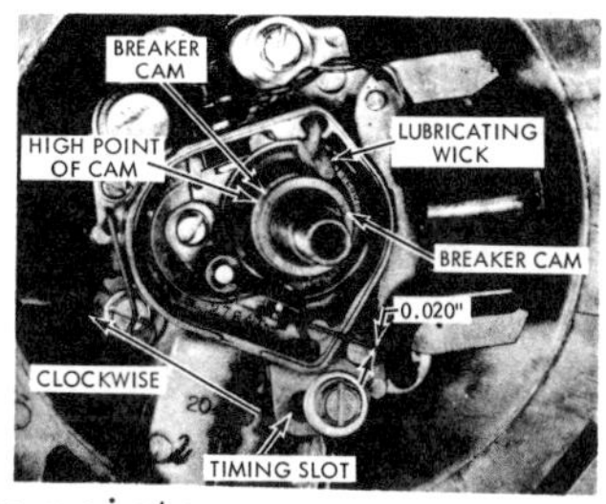

3. Remove the flywheel and inspect breaker points.
4. If pitted or corroded, replace them. See figure 38a.
5. Check all wiring for breaks or short-outs.
6. Check all soldered connections and resolder if needed.
7. Check the coil and condenser. If points are badly burned, replace the condenser also.
8. To set breaker points, turn crankshaft until breaker arm shoe is on high point of cam. Use a feeler gauge and set to .020".
9. To adjust, loosen adjustment screw and move breaker plate to setting shown above.
10. Magneto to edge distance is pre-set at the factory. If the point gap must be changed, and to-edge distance has been distorted, re-gapping the points should also re-establish to-edge distance.
11. Other malfunctions in magneto which may distort to-edge distance are; worn breaker cam, breaker cam with worn key, breaker cam upside down, split breaker cam, worn flywheel or crankshaft keyway, a twisted crankshaft, wide keyways both crankshaft & flywheel, partially sheared key, no torque on flywheel, wide point settings, wrong points, wrong breaker cam, and points in wrong position.
12. When magneto is open, check for oil leakage. If present, check all oil seals and gaskets. Replace if necessary.
13. If the points were pitted, check closely for extent of damage. Pitted or blackened points, and lack of spring tension, require service.
    a. Many mechanics dress pitted points, but it's next to impossible to build up the hole properly. Replacement is the best method.
    b. If the points have a grayish cast, the balance between coil and condenser is good, and points don't require replacing.
14. When replacing breaker cam, always place grease on the crankshaft.
15. If complete overhaul is required, flywheel must be removed. This operation is fairly basic and complete instructions relating to the Jacobsen Model 321 will be found in Chapter Six, on pages 89 thru 93.

Troubleshooting Guide for Engines:

Engine fails to start, or starts hard.

1. No fuel in the tank.	1. Fill tank with clean, fresh fuel.
2. Fuel shut-off valve closed.	2. Open fuel shut off valve.
3. Obstructed fuel line.	3. Clean, or replace with new line.
4. Water, foreign liquid, in fuel.	4. Drain the tank. Clean fuel lines and clean the carburetor.
5. Engine overchoked.	5. Close fuel shut-off and activate starter until engine starts. Reopen fuel shut-off valve.
6. Incorrect carburetor adjust-	6. Adjust the carburetor.
7. Loose magneto wire.	7. Check wire for correct setting.
8. Faulty magneto.	8. Check timing, point gap.
9. Spark plug fouled or bad.	9. Clean, regap plug, or replace.
10. Stop device in OFF position.	10. Move to ON position.
11. Engine flooded.	11. Open choke. Remove air cleaner and clean as needed.
12. Choke valve not closing in carburetor.	12. Adjust control cable travel, and check speed control lever.
13. Carburetor idle needle or the power needle not adjusted.	13. Reset idle and high speed needle at recommended setting.
14. Carburetor throttle lever not open far enough.	14. Move speed control lever to fast position; check for linkage binding, or an unhooked governor hook.
15. Low, or no compression.	15. Check each of the following: a. Blown head gasket. b. Damaged or worn cylinder. c. Valves stuck open, burned, or not properly adjusted.
16. Oil seals leaking(2 stroke).	16. Replace oil seals.

Engine Lacks power.

1. Choke not completely open.	1. Open choke.
2. Improper carburetor adjustment.	2. Adjust as needed.
3. Magneto improperly timed.	3. Time magneto as needed.
4. Worn piston or rings.	4. Replace piston or rings as needed.
5. Air cleaner dirty or plugged.	5. Clean air cleaner.
6. Valves leaking.	6. Grind the valves.
7. Improper fuel mixture for a two stroke engine only.	7. Drain tank, carburetor, and refill according to manufacturers specs.
8. Not enough oil in crankcase, four cycle engines only.	8. Drain and refill to proper level.
9. Engine needs major overhaul.	9. Take to your local dealer.
10. Too much drag on the driven equipment.	10. Adjust clutch, pulleys, or sprockets as needed.
11. Obstructed exhaust system.	11. Remove the obstruction.
12. Bad muffler.	12. Replace the muffler.
13. Crankcase gaskets or seals leaking.	13. Replace gaskets, or seals.

Engine Misses Under Load.

1. Spark plug fouled.	1. Clean or replace if needed.
2. Improper spark plug gap.	2. Regap the spark plug.
3. Carburetor idle, or power needle not properly adjusted.	3. Reset needles to proper setting.
4. Valves not working properly.	4. Reseat, or grind if necessary.
5. Restricted fuel supply.	5. Clean tank and fuel lines.
6. Pitted magneto points.	6. Clean and lubricate points.
7. Magneto arm sluggish.	7. Clean and lube point breaker arm.
8. Faulty condenser.	8. Check it on a tester. Replace if it's defective.
9. Weak valve spring.	9. Replace valve spring.

Engine Knocks.

1. Carbon in combustion chamber.	1. Remove cylinder head and clean all carbon from head and piston.
2. Loose or worn con rod.	2. Replace connecting rod.
3. Loose flywheel.	3. Tighten flywheel nut to proper torque per makers specs. If unit has a key and keyway-replace if needed.
4. Loose drive pulley blade.	4. Replace-retighten as needed.
5. Worn cylinder.	5. Replace cylinder.
6. Main bearings worn.	6. Have main bearings replaced.

Engine Surges or Runs unevenly.

1. Fuel cap vent holes plugged.	1. Open vent hole, or replace cap.
2. Carburetor throttle linkage or throttle shaft, or butterfly is binding or sticking.	2. Clean, lube, and adjust linkage.
3. Carburetor float level too low.	3. Reset the float level.
4. Restricted fuel to carburetor.	4. Clean tank, lines and inlet needle and seat of carburetor.
5. Idle and power needles improperly adjusted.	5. Readjust power and idle needles.

Engine Overheats.

1. Engine out of time.	1. Have engine timed.
2. Carburetor out of adjustment.	2. Adjust per previous instructions.
3. Air flow obstruction.	3. Remove obstructions from air passages in the shrouds.
4. Overspeeding engine.	4. Reset speed control.
5. Cooling fins clogged.	5. Clean cooling fins.
6. Overloading engine.	6. If this happens often, review need for a larger engine.
7. Not enough oil in crankcase.	7. Drain & refill to required quantity
8. Improper fuel mix(2 cycle).	8. Drain, remix to makers specs.
9. Engine dirty.	9. Clean grease and dirt from block.
10. Carbon in combustion chamber.	10. Remove head and clean as needed.
11. Engine out of time.	11. Have engine re-timed.

Engine Vibrates Excessively.

1. Engine not mounted properly.	1. Tighten mounting bolts.
2. Bent crankshaft.	2. Replace, or have replaced.
3. Blades, adapters, pulleys, & sprockets may be out of balance.	3. Re-align, or replace parts as required to correct.

Spark plug equivalent and heat range chart for most small engines.

	CHAMPION			AUTOLITE			NGK
	Reg.	2-cycle	Racing	Reg.	2-cycle	Racing	
Hot		J14J			A11X		
		J12J			A11X		B4
	J-11	J11J		AT8	A9X		
14MM	J-8	J8J		AT6	A7X		
3/8"		HO-8A					
Reach	J-7	J7J		AT4	A5X		B6
	J-6	J6J		AT3	A3X		
			J62R	AT2		A23	B7
	J-4	J4J		AT1	A21X		
			J57R	A901		A603	B8
			J54R			A403	B9
Cold						A203	B10
Hot	H-12	H12J		AL9			
7/16"	H-11	H11J		ATL8			
Reach	H-10	H10J		ATL4	AL7X		
Cold	H-8	H8J		ATL3	AL5X		
Hot	L-10			AE6			B4H
	L-7			AE4			
1/2" reach			L-62				B6H
	L-5			AE2		AE23	B7H
		L4J	L-57R	AE901		AE603	B8H
			L-54R			AE403	B9H
						AE203	B10H
Cold							
Hot	N-18			AG7			B4E
	N-8			AG5			
							B6E
	N-5		N-62R	AG4		AG23	
3/4" reach				AG3			B7E
	N-3			AG2			
				AG901			
			N-57R	AG701		AG603	B8E
			N-54R			AG403	B9E
						AG203	B10E
Cold							

Figure 37

Figure 38

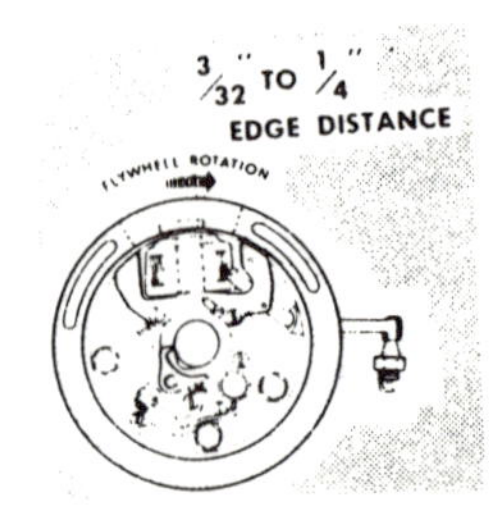

Figure 38a

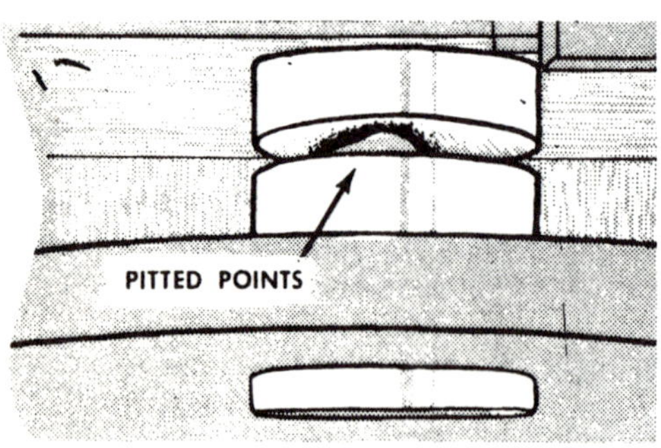

Figure 38b

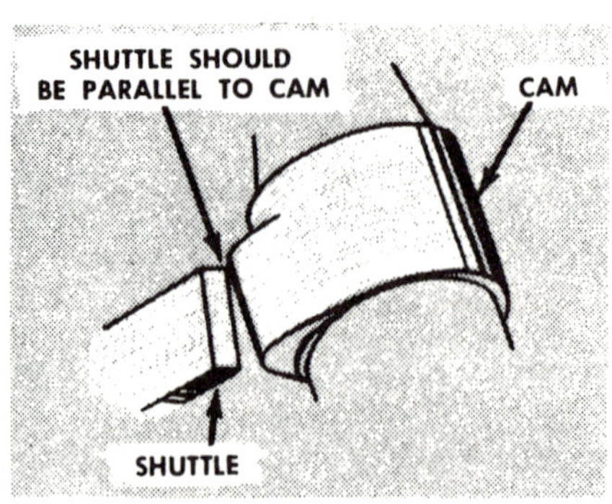

Chapter Three

Pioneer Gen-E-Motor Corporation(Pincor)

Pincor ceased manufacturing engines around 1965, but many are still in operation so the following coverage will refer to that engine.

Preparing the Engine for operation.

1. Check oil level, and change oil after 25 hours of operation.
2. Open oil fill plug. Use pliers or adjustable wrench.
3. Engine should be level. Place mower on driveway.
4. Pour oil into fill hole(fig. 39), until level with top of hole.
   a. Use SAE 30 oil.
   b. Capacity is 1 1/4 pints.
5. Add oil to the air cleaner at the same time.
   a. Remove wing nut from top of air cleaner.
   b. Remove air cleaner element.
   c. Pour oil into bowl to level marking(fig. 40).
   d. Use SAE 30 oil.
   e. Replace element in bowl. Don't force it into place. Forcing it could damage the element and reduce filtering ability.
   f. Fasten element to bowl with wing nut.
6. Fill fuel tank with regular gasoline.
   a. NEVER FILL TANK WHILE ENGINE IS RUNNING.

Starting the Engine.

1. Open shut-off valve by turning outward(fig. 42).
2. Always close the valve when mower is not in use for prolonged periods of time, such as over a winter, or during a long rainy spell.

Figure 39

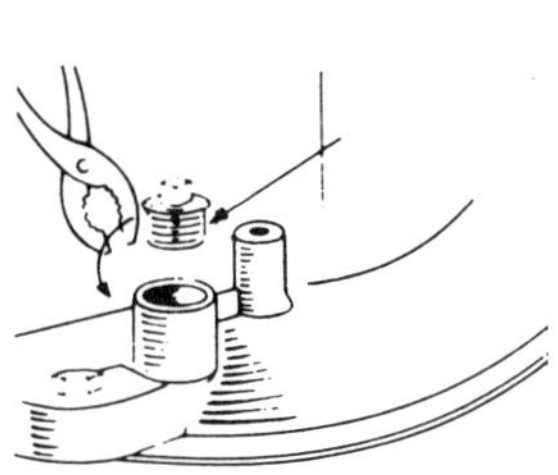

Figure 40

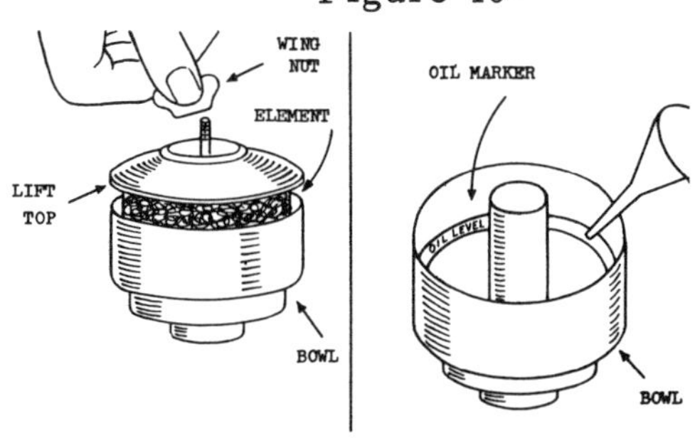

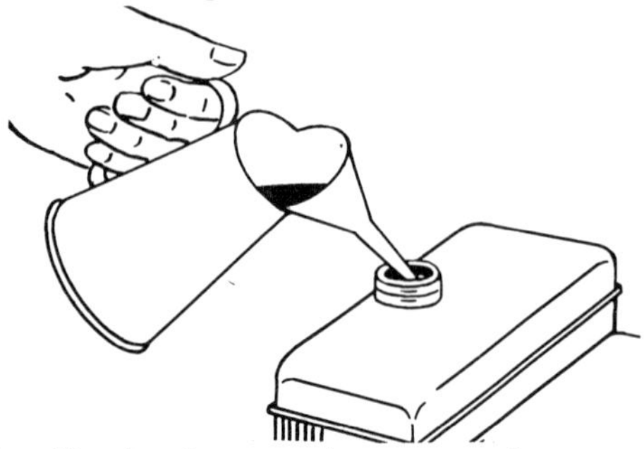
Figure 41

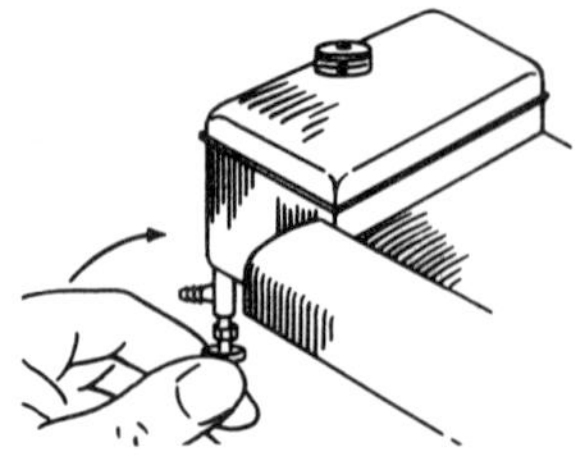
Figure 42

3. Push stop spring away from spark plug.
4. Move choke lever to "UP" position. "UP" is full choke. "DOWN" is open choke position.
5. Starting procedure.
   a. If mower is equipped with rope starter, turn engine counterclockwise until you feel compression. Slip rope into notch on the pulley located at top of engine. Wind rope around, clockwise, until it's all on the pulley. Pull rope briskly until all rope unwinds. Repeat this procedure until the engine starts.
   b. If mower is equipped with a recoil starter, pull rope out slowly until the engine rolls past compression stroke. Allow automatic retriever to retrieve rope until about 2" remain exposed. Pull the rope briskly until the engine starts.
   c. If mower is equipped with a snap starter, unfold handle and turn 1/2 cycle. Place locking lever in "LOCK" position. Turn handle clockwise 4 times. Fold handle and stand back away from mower. Release lever and mower should start. Repeat as needed.
   d. If crank starter won't start engine, starter may not be wound enough, or the mower blade may be jammed. Rewind starter per above instructions and try again.
      A. If mower blade is jammed, push lock lever to "LOCK", and disconnect the spark plug wire. Check blade for excess grass. Clean underside of mower as needed.
      B. If you disassemble the snap starter, don't remove the spring from steel housing. The spring is extremely powerful. If released it could cause serious bodily injury.
   e. After engine starts, follow common procedure and move choke lever back to open position.
6. To stop engine, push STOP spring against the spark plug.

Figure 43

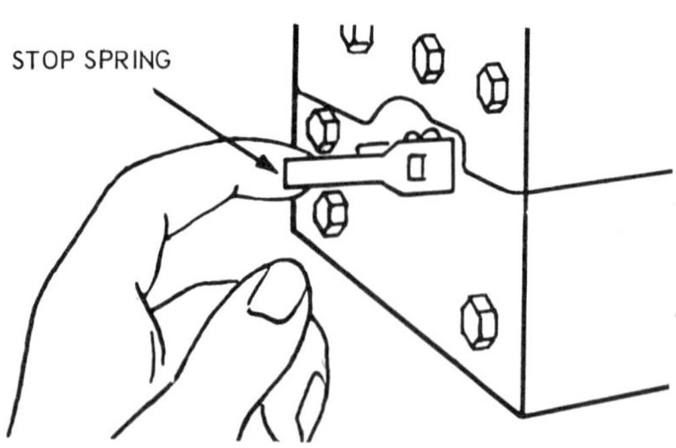

Figure 44

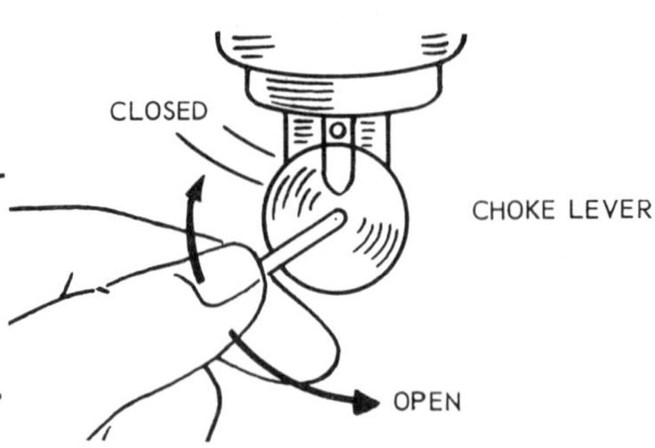

Figure 45

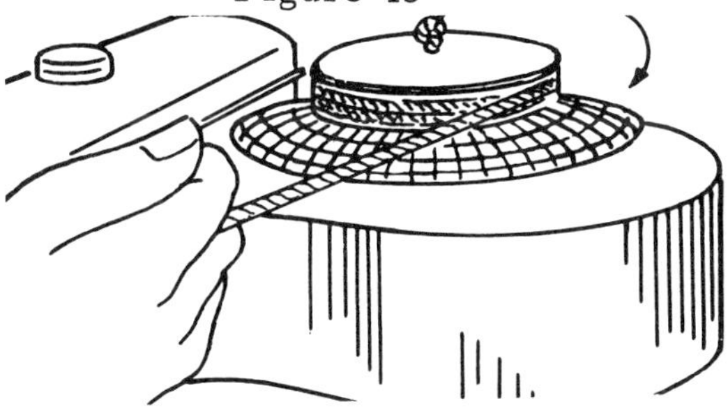

Figure 46

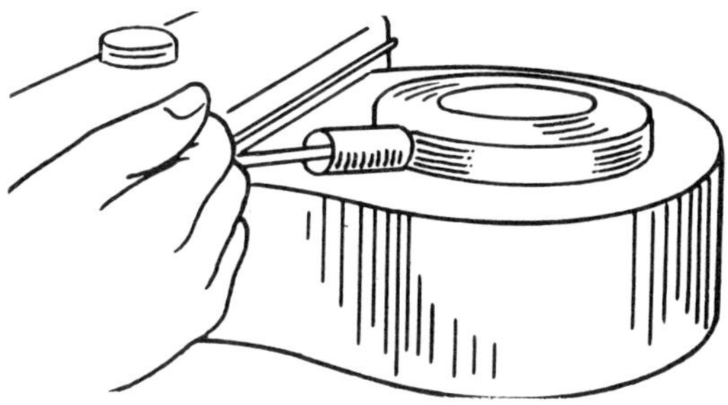

General Maintenance and Service.

1. Change oil after 25 hours of operation. If required, add fresh oil every 5 operating hours.
2. Oil drain plug is located under the mower. See figure 47.
3. Place mower off the floor in a level position. Remove drain plug and allow oil to drain into a receptacle. Always change oil when engine is warm. The oil will drain faster.
4. Clean and refill the air cleaner every 10 hours, more frequently if dusty conditions prevail.
5. Remove wing nut from top of element and lift element out.
6. Pour out old oil and rinse the bowl with gasoline.
7. Clean the element by dunking in gasoline a few times. Allow several minutes for element to dry. After element is dry, dunk in oil and shake out the excess oil.
8. Refill bowl to marked oil level and replace element. Tighten wing nut.

The Spark Plug.

1. Spark plug should be cleaned and re-gapped occasionally.
2. Gap setting is approximately .025". Use a feeler gauge if possible.
3. If replacement is required, use Champion J-8 or equal.
4. To clean, scrape away corrosion and carbon. Refer to page 30, in Chapter Two for additional spark plug service tips.
5. To check engine for spark, remove ignition cable from plug.
6. Place cable tip 1/8" from cylinder head bolt.
7. Turn flywheel rapidly for several turns.
8. If spark jumps the gap, ignition is in order and spark plug must be replaced.
9. If spark doesn't jump, refer to chapter on engine repair and follow steps covering that phase of operation. Different engines vary somewhat, but basic service procedures apply in general service.

Figure 47

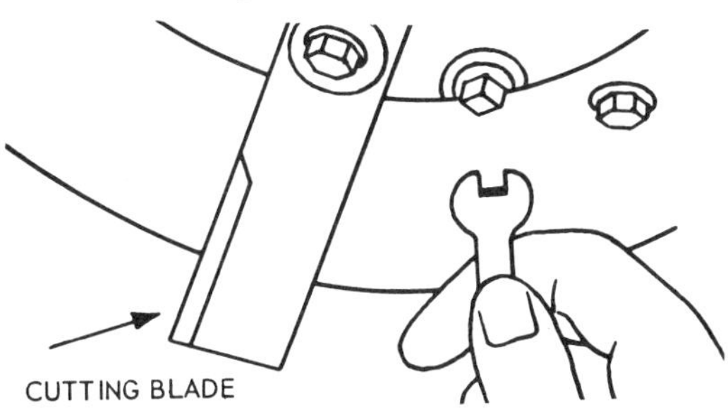

Figure 48

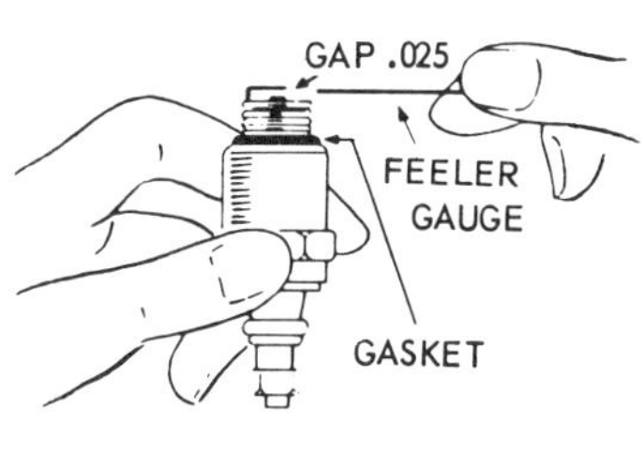

Figure 49

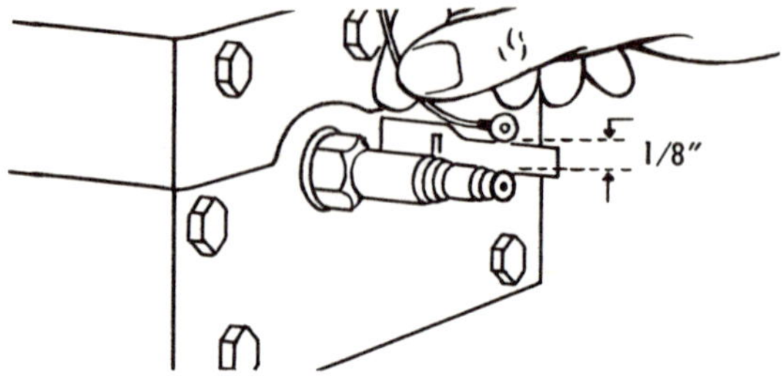

Carburetor Adjustment.

1. To adjust carburetor illustrated in figure 50, open high speed adjusting screw and idle jet screw about 2 1/2 turns from closed position. Do not force jet screws. If screws won't turn, apply some penetrating oil. Allow oil a few minutes to soak in and try again.
2. Start engine and allow it to run for several minutes until reaching normal operating temperature.
3. Set throttle approximately 3/4 open by moving governor lever to the left.
4. Turn high speed jet screw in, until engine slows or coughs.
5. Note the position, and turn screw out until engine slows down or runs unevenly.
6. The correct adjustment should be between these two points, or about 1 1/2 turns open.
7. Set throttle to closed position by moving governor lever to the right.
8. Adjust idle jet by turning inward slowly until engine idles roughly.
9. Back off 1/4 to 1/2 turn until engine idles smoothly.
10. Idle jet screw should be open approximately 1-1/2 turns.
11. Hold throttle in closed position and check idling speed.
12. If engine idles too fast, turn idle speed regulating screw outward.
13. If engine idles too slow, turn regulating screw in as needed.
14. If additional adjustment to idling circuit is required, repeat steps 7 through 13, until adjustment is correct.
15. Don't allow engine to idle below 2000 RPM's.
16. Don't run engine above 3600 RPM's.

Carburetor Adjustment for figure 50 .

1. Make carburetor adjustment with fuel tank half filled with gasoline.
2. Close needle valve(figure 50) clockwise.
3. Turn it out, counterclockwise, 1 1/2 turns and start engine. After a few minutes, engine will warm up to normal operating temperature.
4. Adjust throttle until engine is turning about 3000 RPM's(no load).
5. Turn needle valve clockwise(inward) until engine loses speed.
6. Reverse the step and turn needle valve counterclockwise, pass the point of smooth operation until engine starts running unevenly again. At that point the engine should perform well under load conditions.

Figure 50

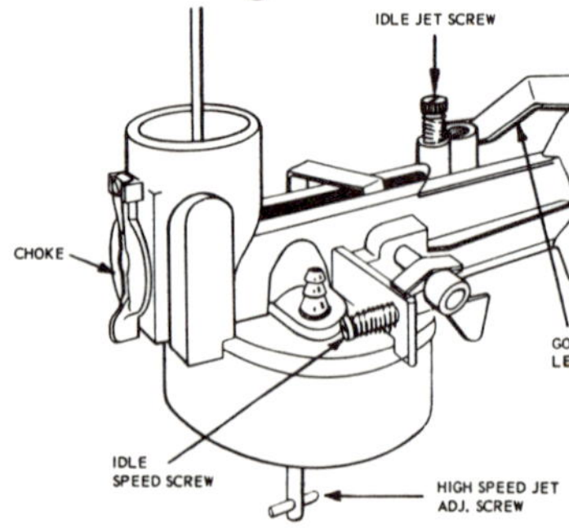

Figure 50a

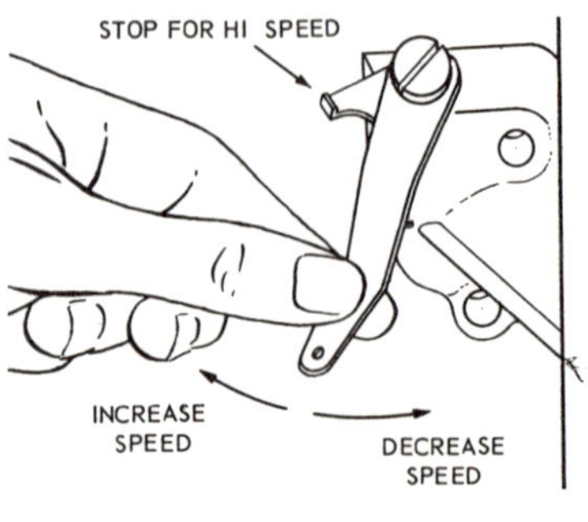

7. To adjust idling speed, push throttle against idle position speed adjusting screw behind carburetor.
8. To increase idle speed, turn adjusting screw inward.
9. To decrease idle speed, turn adjusting screw outward.
10. If engine has a tendency to stall or die out under load(while cutting), the mixture is probably too lean. Refer to steps above. If engine runs unevenly during idle, mixture is probably too rich. Adjust per above.

Adjusting Engine Speed.

1. To adjust engine speed, it's necessary to re-position governor lever.
2. To lower engine speed, move lever toward carburetor.
3. To increase engine speed, move lever away from carburetor.
4. To adjust the highest no load speed, loosen screw on lever (fig.50a and rotate spring washer stop away from carburetor as needed, and re-tighten governor lever screw securely.
5. To lower the no load speed, loosen screw on lever and rotate spring washer stop toward carburetor. Re-tighten governor lever screw.
6. The throttle and governor unit should always work freely. If any binding occurs, clean the area, inspect linkage points and adjust when necessary.

Control System.

1. To start engine on mowers equipped with the model 7V or 7VR, move lever to right in CHOKE position.
2. Move choke lever on carburetor to UP position.
3. Crank engine to start.
4. If mower fails to start after several tries, move choke lever back toward OPEN position. Avoid overchoking.
5. After engine starts, move lever to desired speed.
6. To service control knob, remove air cleaner from carburetor.
   a. Remove 2 screws holding control to carburetor.
   b. Disconnect governor spring from lever and remove the lead wire from underside, by removing round head screw.
7. To assemble, reverse the above procedure.
8. To stop engine, push control to STOP position.

The Recoil Starter.

The recoil starter on Pincor engines automatically rewinds the starter rope. If the spring loses tension, or the rope is worn and needs replacing, proceed as follows:

1. To replace the rope, remove 4 retaining screws (71-044, fig. 51b), holding cover(71-047) to engine, and remove the starter assembly.
2. Place assembly on work bench with mechanism up, and remove the screw(71-058).
3. Remove washer(71-052), spring(71-059), and pawl retainer(71-055) from the unit.
4. Rope knot is now exposed. Remove the knotted end with a pair of pliers. Place a wedge against pulley(71-053) so rewind spring(71-057) won't cause it to turn when knot is released. Ease the wedge out and allow pulley to unwind slowly.
5. Before replacing the rope, wind rope pulley counterclockwise until spring is tight. Back off two turns, allowing hole in rope pulley to align with guide bushing. Wedge the pulley again.

Figure 51

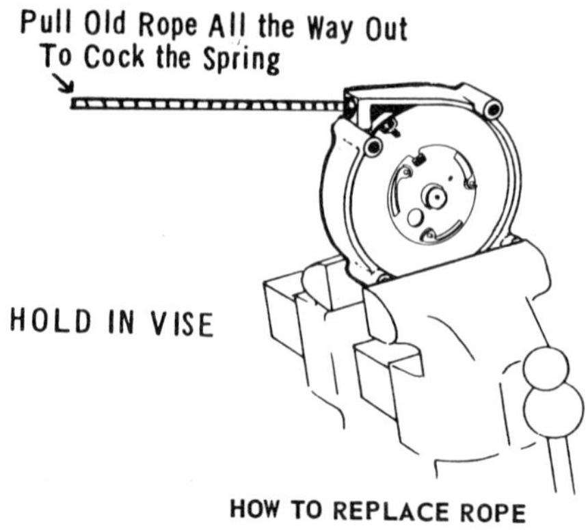

Figure 51a

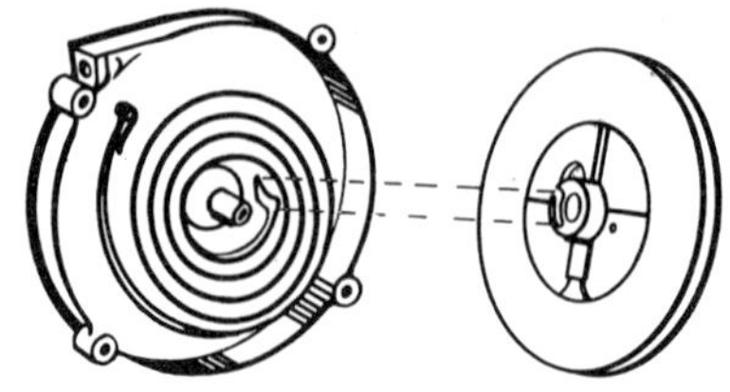

6. Hold rope pulley in position and insert new rope through guide bushing and through rope pulley. Tie a single loop knot, with short end of rope no more than 1/4" long. Coat the knot with shellac. It will hold longer and set in place faster.
7. Hold rope and remove wedge, allowing pulley to retrieve the rope slowly.
   a. Pull rope briskly several times to check recoiling action.
8. Re-assemble unit in reverse order. If a vise is available, use in manner illustrated in figure 51 .
9. Remember, the recoil spring is powerful. Use extreme caution when winding the pulley to full tension, and don't allow it to slip when winding.
10. To replace recoil spring, follow above steps and remove the rope.
11. The spring is exposed when rope is removed. Start from the center and allow spring to uncoil itself slowly as you bring it out.
12. To replace the spring, hook loop to post of housing. Figure 51a.
13. Wind spring in place by turning housing clockwise and feeding spring in slowly. Use caution. Spring could snap out of position and cause a serious injury.
14. Replace rope pulley in housing. Hook spring around hub of pulley by guiding spring into slot of pulley with a small screwdriver.
15. Replace other components per previous instructions.

Figure 51 b

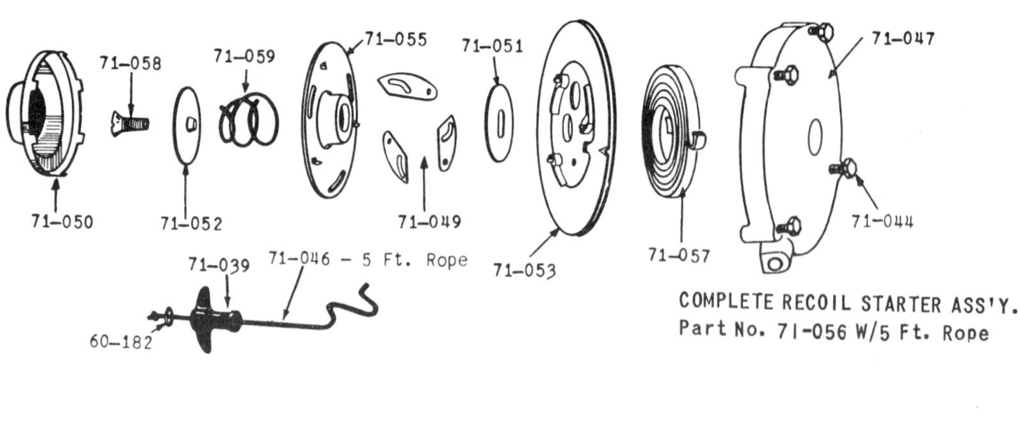

Figure 51 c

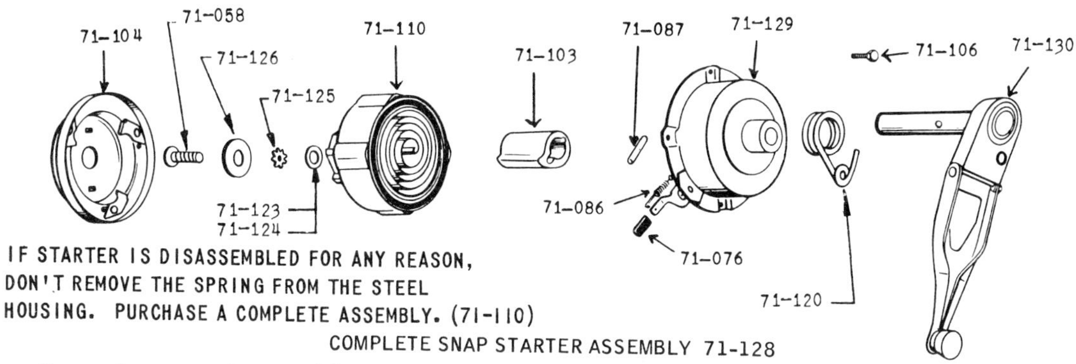

Snap Starter Assembly.

1. Service to this unit generally applies to main spring, which is housed in unit(71-110). If the spring breaks, or loses tension, replace the entire unit. Never attempt changing the spring. Never remove it from the housing. Serious injury could result. See figure 51 c above.

Figure 52
Parts breakdown on carburetor assembly 60-1684

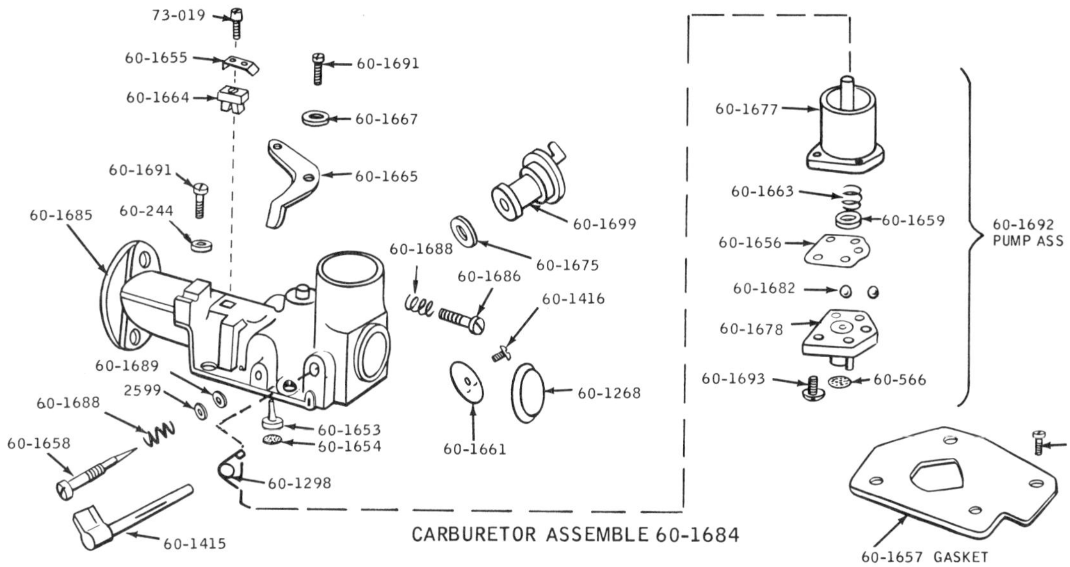

COMPLETE PARTS LIST

t No.	NAME OF PART	Req.	Part No.	NAME OF PART	Req.
76	Washer	6	60-242	Oil Fill Plug	1
79	Key	1	60-244	Washer	1
99	Washer	1	60-566	Screw	
-1620	Exhaust Valve	1	60-620	Muffler	
-023	Valve Springs	2	60-644	Oil Seal	1
-044	Plain Washer	4	60-745	Breaker Point Set	1
-061	Spark Plug & Washer		60-1001	Push Rod	2
-092	Screw	1	60-1003	Piston Pin	1
-199	Locknut	1	60-1011	Retainer for Piston Pin	2
-239	Nut	1	60-1024	Large Gear (used with 60-1271)	1

Figure 53
Component Parts - Pincor Engine.

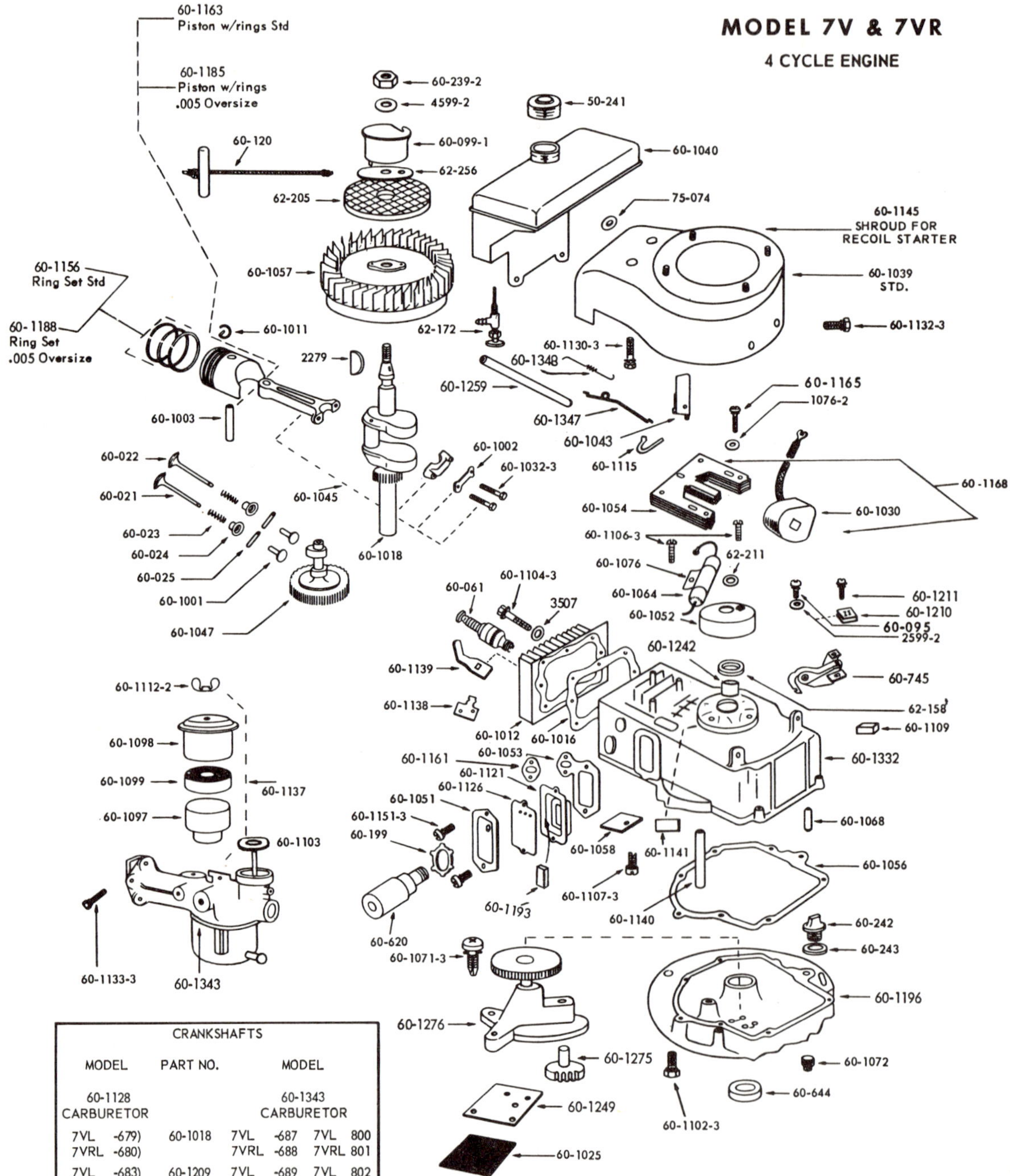

CRANKSHAFTS

MODEL	PART NO.	MODEL	
60-1128 CARBURETOR		60-1343 CARBURETOR	
7VL -679)	60-1018	7VL -687	7VL 800
7VRL -680)		7VRL -688	7VRL 801
7VL -683)	60-1209	7VL -689	7VL 802
7VRL -682)		7VRL -690	7VRL 803
7VL -684)	60-1212	7LV -691	7VL 804
7VRL -685)		7VRL -692	7VRL 805
7VL -686)	60-1213	7VL -693	7VL 806
7VRL -681)		7VRL -694	7VRL 807

ENGINE GASKET SET - 60-1162

Service for Reel Type Power Mower

Figure 54

Parts breakdown on Model 5518, 18" Reel type, self-propelled mower.

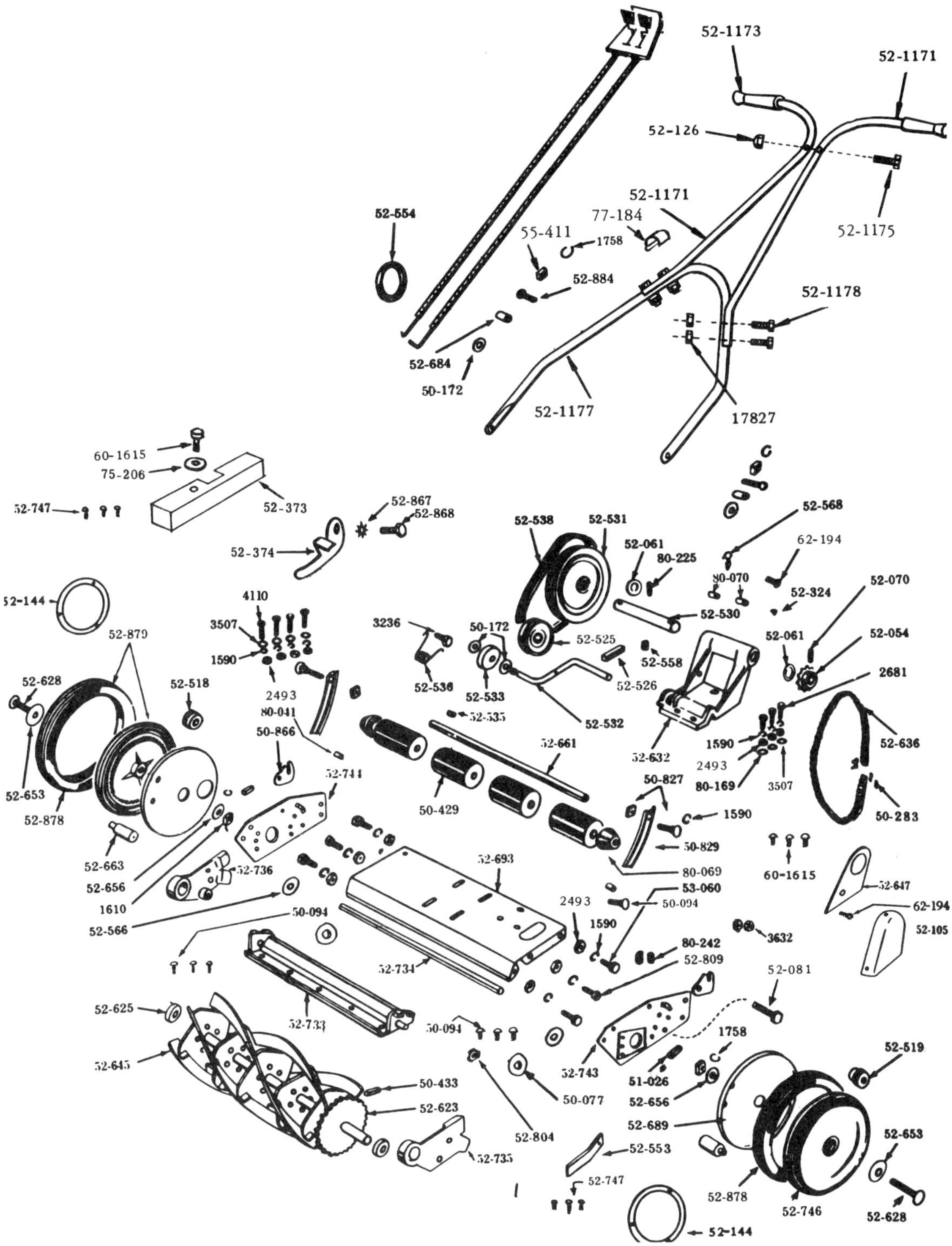

Engine service and starting procedures have been covered at the beginning of the chapter. Remember, the mower is self propelled. When engine starts, pull throttle back to the SLOW position until ready to move.

When you're ready to move, push clutch lever to ON position. To stop the mower, move clutch lever back to STOP position.

Blade Adjustment.

1. The blade adjustment is correct when the reel blades just touch cutter blade evenly from end to end.
2. If adjustment is required, loosen the two slotted screws located in each side of the side plates.
   a. The top screws adjust the cutting blade away from the reel, and the lower screws adjust cutting blade toward the reel.
   b. After adjustment is made, test with a sheet of newspaper. Slip newspaper in between blades. When blade just touches, adjustment is correct.
   c. Or adjust with the newspaper. Loosen top screws and tighten bottom screw until cutter blade just holds the paper, use right side first. Repeat with left side adjustment. When the paper is held at both ends, and the reel touches cutter evenly at all points, while top adjusting screws are tight, the unit is properly adjusted. Tighten locknuts on lower screws to complete adjustment.

Cutting Height Adjustment.

1. Most mowers are factory set at about 1 1/2", but are adjustable from 1/2" to 2 1/2". The side plate(52-743-744) is equipped with three axle positions.
2. When the wheel and axle are in the upper hole, the cutting height is approximately 1/2". Moving the wheel and axle to the middle hole will give factory setting of 1 1/2". The third hole adjustment will make a cutting height of 2 1/2", all with a level top plate. The in between adjustments can be made, but engine operates most efficiently when the rear of top plate is within 1/2" above, or below, a level position.
3. If completely different cutting heights are required, loosen the hanger bolts(827) and move rollers(50-429) upward evenly to lower the cutting height, or downward to raise the cutting height.
4. After adjustment is made, re-tighten the roller hanger bolts.
5. To change wheel position, remove axle bolt, axle dust guard, and a wheel, intact, by removing nut on other side of side plate. Place the entire assembly in desired hole and tighten bolt. Always position both of the wheels in same hole.
6. When making roller adjustments, set mower on level surface so both brackets are set for equal height.

Sharpening the Cutting Reel.

1. Disengage the clutch control. Remove motor drive V-belt from the jackshaft pulley(14) figure 54, and also from engine pulley.
2. Place round rubber sharpening belt(a) on jackshaft pulley.
3. Twist belt once and place other end on motor drive pulley. Fig. 55.
4. Start engine. Twisting the belt causes reel to turn in reverse, and sharpens the cutting blades.
5. Use a small brush, and apply a mixture of emery material and oil evenly along entire length of reel. Continue until a rubbing noise disappears completely. It should require about 5 minutes.

If blades are very dull, tighten the reel and cutter adjustment, and repeat the procedure.

6. Stop the engine and wash emery and oil from area with gasoline.
7. Remove sharpening belt, and replace regular motor V-belt.

Figure 55

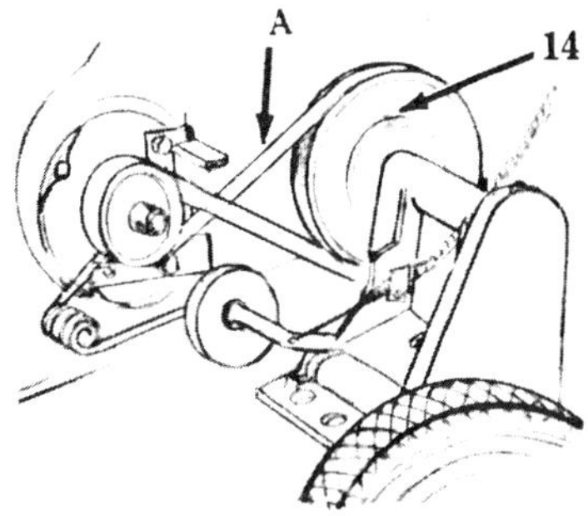

Sharpening the Cutting Reel

Cleaning Wheel Gear and Pinion.

1. To clean wheels, remove them by removing bolt(52-628) and washer holding wheel to chassis, then removing the wheel.
2. Remove pinions (52-519), (52-518), and pawls(51-026).
3. Clean all grease and dirt away from wheels and pinions. Use gas or kerosene.
4. Replace all parts in reverse order of removal.
5. Repack pinion with light grease and replace on end of shaft. Replace pinions on proper side.
6. Check for proper setting by a turning of pinions on shaft forward as if wheel was turning forward. If pinion turns forward freely, but it locks when turned in opposite direction, replacement is correct.
7. Place grease on wheel shaft and replace wheels. Replace washers, and bolts, holding wheel to chassis, and tighten securely.
8. Oil area between wheel washer and wheel, between washer and idler pulley, and in oil cup on top of jackshaft bracket, especially during the mowing season.

Figure 56

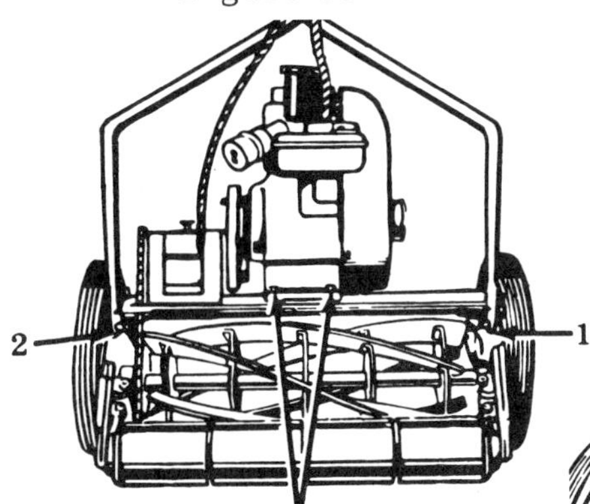

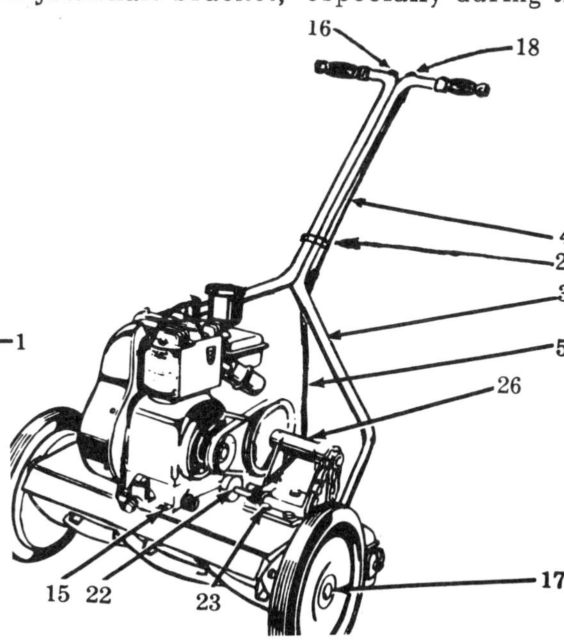

Adjusting the Controls.

1. Throttle control line should be properly attached to the engine.
2. Set throttle control at handle on FAST speed setting.
3. Loosen control casing clamp screw(B) on carburetor. Fig. 58
4. Move control casing(A)and wire proper direction to enable speed lever(C) to just touch choke operating link at point(D).
5. Tighten casing screw(B).
6. Recheck control after adjustment. When throttle lever is pulled back, it must activate full choke. Engine should stop when lever is pushed completely forward.

Figure 58

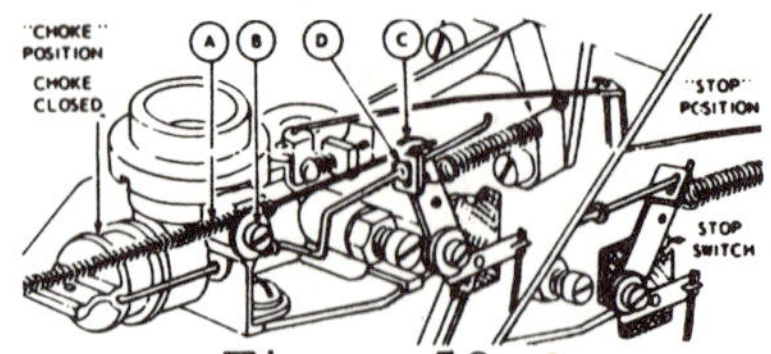

Figure 59

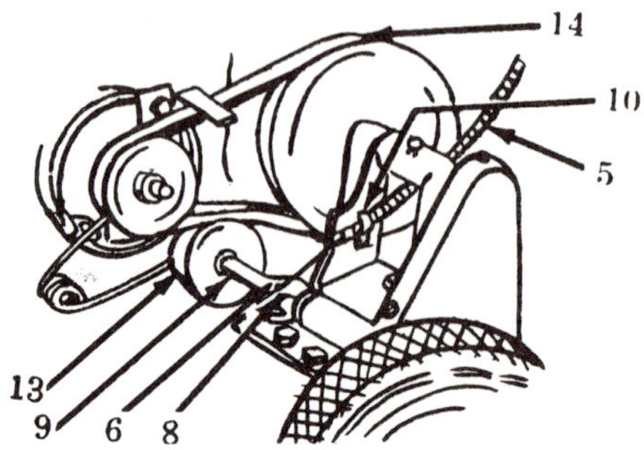

Clutch Control

1. Clutch control(18)fig. 57, must be properly adjusted and give a positive snap when lever is pushed forward to engage idler.
   a. When lever is moved back, it should disengage idler.
2. To adjust, remove bolt-4110, figure 60, and mower cover.
3. Bent portion of clutch control wire(6) fig. 59, should be inserted in hole on idler arm(8).
4. Loosen cable clips(10). Insert cable casing under clip groove, and allow about 3/4" of casing to extend past the clip. Tighten clip securely.
5. If control works sluggishly, drop some oil into each end of casing.

Chain Adjustment.

1. After stiffness is worked out, chain will stretch somewhat.
2. When chain becomes loose and sags, adjustment is required.
3. Loosen bolt(23) and move jackshaft toward handle end, until chain is snug, with about 1/8" of lateral play.
4. Hold firmly in place and tighten bolts(23). The chassis is slotted to allow adjustment.
5. Before final tightening, check jackshaft position. It must be set exactly parallel with edge of top plate.
6. Always double check all nuts and bolts for tightness. A few loose nuts and bolts may result in future serious damage.

Belt Adjustment.

1. Most belts have a tendency to stretch during first few hours of operation, and continue to do so under prolonged use.
2. To adjust, loosen bolts(15), and move engine as needed. Adjustment is correct when idler (13) rests on the belt about 3/4" away from large pulley(14).
   a. Never move the jackshaft to adjust belt.

Bearings.

1. Keep all bearings greased, and all moving parts properly oiled.
2. Oil jackshaft at points(22 & 26), figure 57.
3. Oil each wheel at point(17).

Figure 60
Parts breakdown on Pincor Model 5521, 21" Reel Lawn Mower.

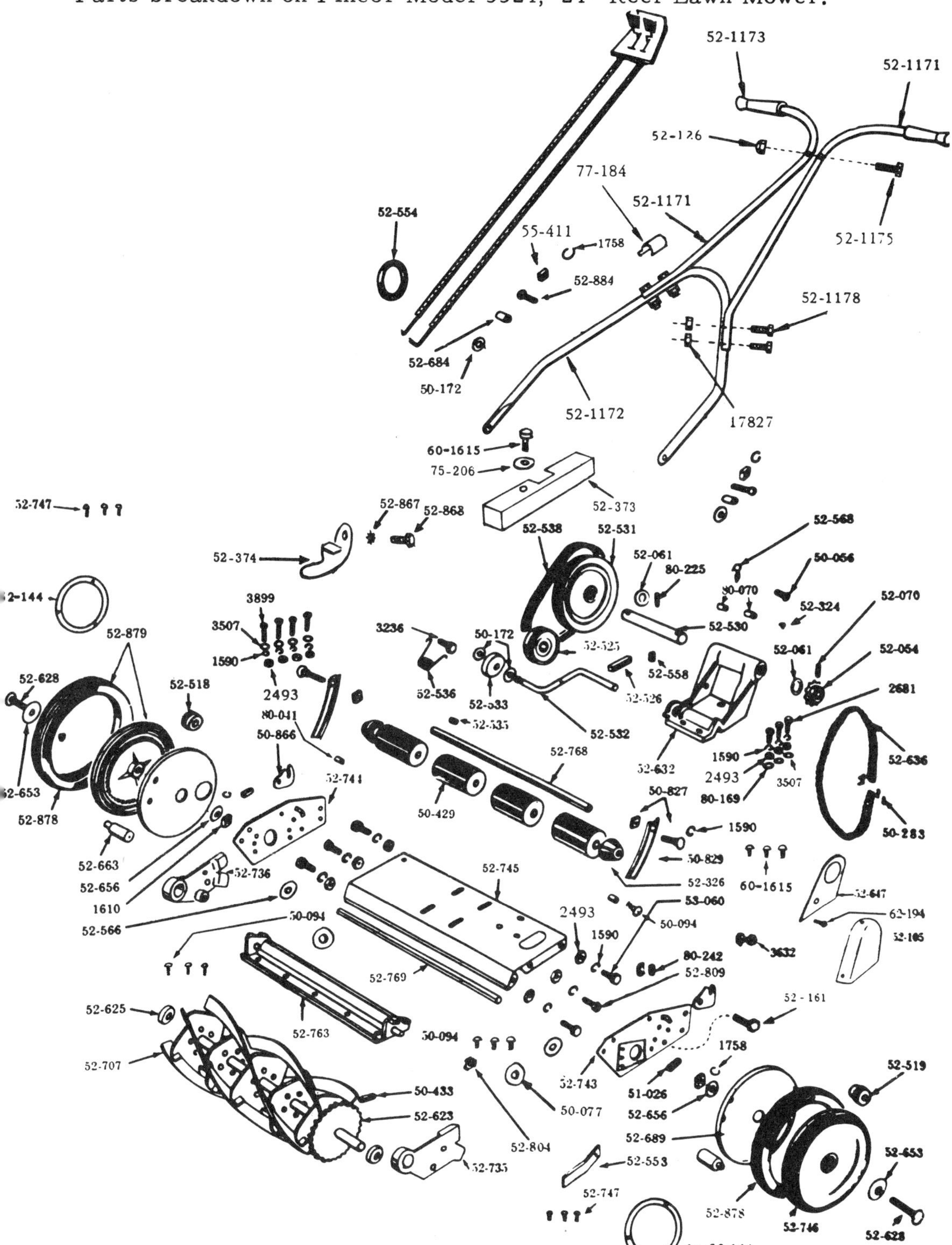

Pincor Rotary Lawn Mowers, Models 7G, 7H, 7HG, 7HSG, 7VHG

It can't be emphasized enough. Always read your instruction manuals before attempting service work. Although most reel mowers are basically alike, and most rotary mowers are basically alike, each manufacturer incorporates different innovations, which require different service procedures.

Read your warranty and don't violate terms if free service time hasn't elapsed.

Assembly.

1. Assemble handle with required nuts and bolts(fig. 61).
2. Attach handle to mower by slipping hole over stud(see insert), and adding the retaining washer and clip.
3. For fold down handle operation, press handle plate away from center of mower.
4. Swing handle over engine until it rests on mower.
5. Loosen bolt wing nuts until they stop against lock nut.
6. Swing top of handle back over engine, but don't kink control cable.
7. Reverse procedure to place handle back in operating position.

Control Assembly.

1. To properly set control lever for operation with Briggs and Stratton engine, first remove the air cleaner. See figure 62.
2. Mount screw and cable clamp to engine shroud. See figure 62
3. Insert wire into hole at point(F).
4. Slip casing(B) under clamp. Do not tighten at this time.
5. Set control lever(E) on control head in FAST position.
6. Adjust lever(C) until it just touches boss(D).
7. Fasten screw (A) securely with knob (E), in FAST position, and the lever(C) touching boss(D).
8. Check setting by moving control lever(E) back and forth, with a final setting in FAST position. Lever(C) should touch boss(D).
9. Clip throttle cable to handle with cable clip.
10. Start engine and try control at various settings. Stop setting should stop engine immediately.
11. To properly set control lever for operation with Tecumseh engine, see figure 63.
12. Mount screw and cable clamp to engine shroud.

Figure 61

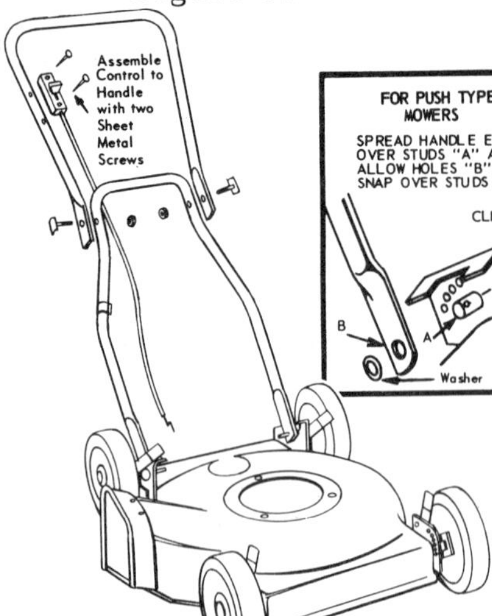

Figure 62

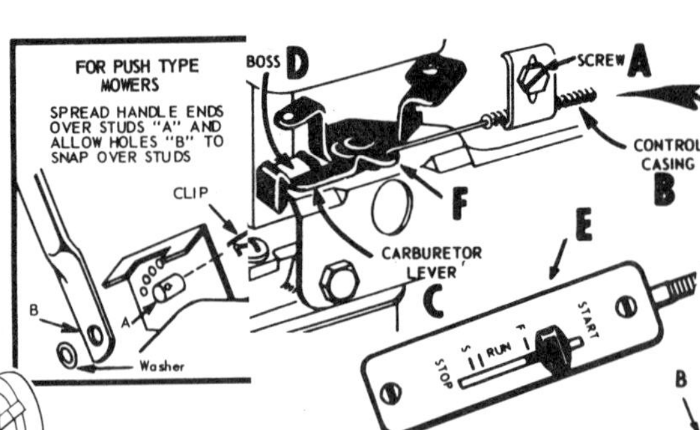

Figure 63

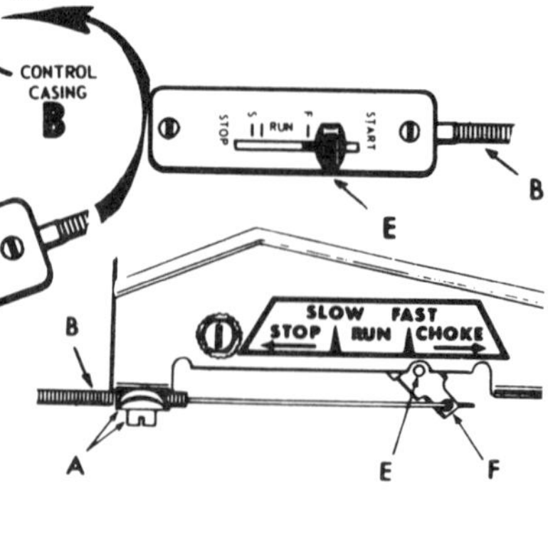

13. Insert wire into hole at point(F), figure 63.
14. Slip casing(B) under clamp. Do not tighten at this time.
15. Set lever(E) on control head in FAST position.
16. Fasten screw (A) securely while holding lever(E) in FAST position.
17. Check setting by pushing control knob(E), located on control panel, into START position. This position should place carburetor on choke.
    a. To be certain, look into intake on carburetor.
18. If choke covers intake, adjustment is correct. Clip cable in place.

Maintenance.

1. Maintenance procedure is similiar to previous steps covering reel-type mowers.
2. Follow instructions covering oil changing, air cleaner service, operating instructions and storage care.

Figure 64

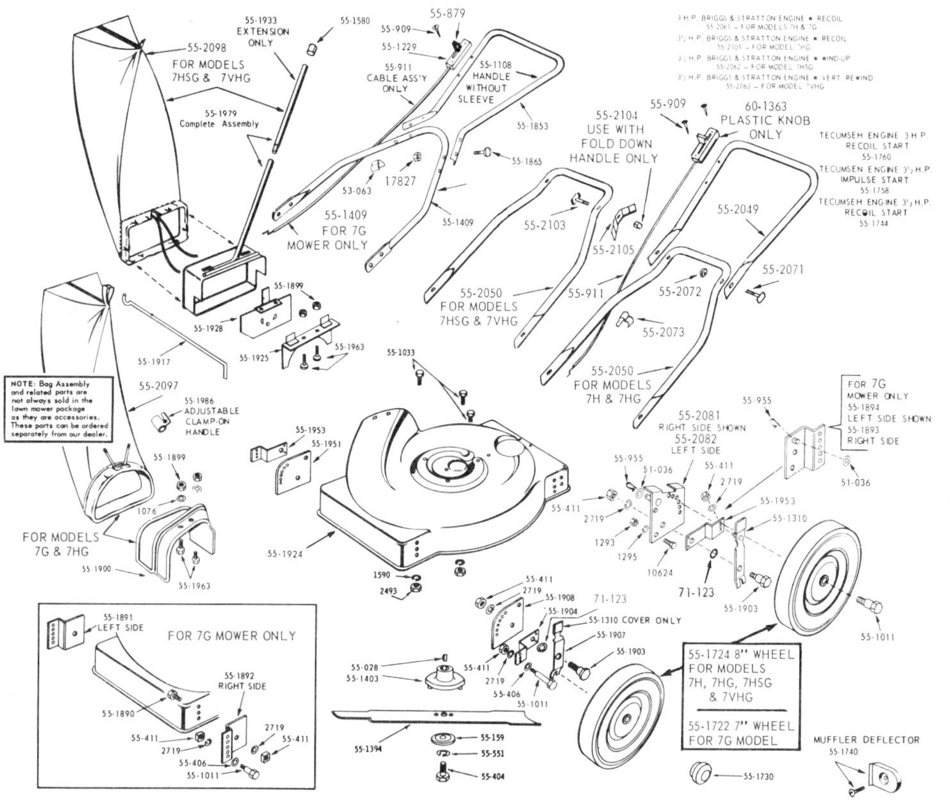

3. When cleaning underside of mower, always disconnect spark plug wire, and remove the air cleaner.
4. When cleaning mower, always tilt unit forward so oil will drain to bottom side of engine.
5. If oil is allowed to drain toward the spark plug, combustion chamber will become oil soaked and engine will not start.
6. Clean underside of mower frequently, it'll increase cutting efficiency and do a better all-around job.
7. Oil wheel/axle and wheel plate mechanism frequently.
8. Don't allow mud and grass to accumulate around driver and wheel-plate mechanism.
9. Use your garden hose to clean away excess grit.
10. If your mower is so equipped, check gear box and grease at least, oftener is preferred, once a season.
11. Check for loose nuts and bolts frequently.

Height Adjustment.
1. If cutting height adjustment is necessary, see figure 65.
2. Increments are shown for both front and rear wheel. Adjust as required, but be certain all wheels are adjusted equally. Incorrect adjustment will cause uneven cutting.
3. To change setting, pull spring lever against rubber tire and rotate wheel and spring lever, together, to desired setting. When changing setting on rear wheels, lift mower off the ground.
4. Always stop engine before making the adjustment.

Figure 65

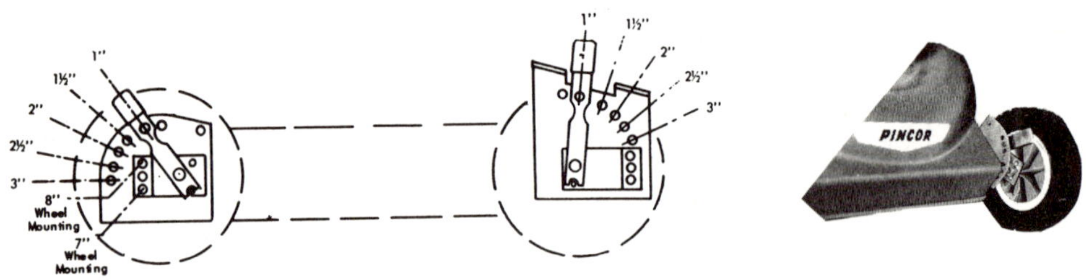

FRONT WHEEL

Blade Service.
1. A sharp blade always cuts better than a dull one.
2. Always remove blade from mower to sharpen or service.
3. Disconnect spark plug wire.
4. Tilt mower forward. Hold blade firmly and remove nut.
5. If blade is badly bent and completely out of balance, replace it.
6. If sharpening is required, use a quality grindstone and sharpen it in several stages, equally on each side. Don't oversharpen either end.
   a. Check frequently for proper balance. Pound a nail into the wall and place blade on it. Rotate blade to find heavy end. Sharpen until proper balance is attained.
   b. Never replace a blade until properly balanced. The vibration could cause irreparable harm to the engine.
7. If you're unsure of the job, have blade sharpened by a professional.
8. When replacing blade, replace all washers, or other components you removed from the engine shaft.
9. Tighten replacement nut securely.

Models 8S, 8V, 28V, and 28BH.

Models in the 28 series are combination self-propelled and push-type units. Lift the handle and the mower is propelled forward by engine power and when the handle is lowered, the power drive is disengaged.

Models 8S and 8V are not self-propelled units but other service adjustments are similiar, including the controls.

Handle assembly procedures are different on each series.

Figure 66
Handle assembly for Model 8 series.

Figure 67
Handle assembly for 28 series.

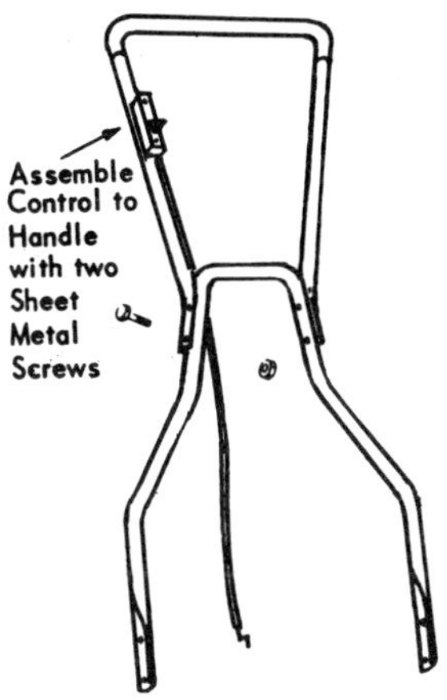

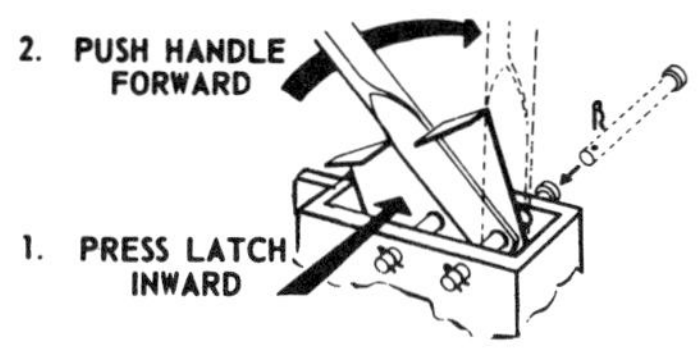

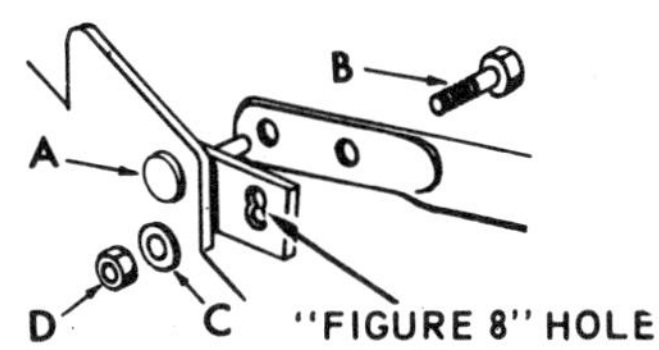

Gear Box Adjustment.

1. Adjustment is necessary when excessive racking of gear box occurs.
2. Loosen locknut (C) and adjust screw (B) downward until it contacts cover(A).
3. Hold screw(B) in place with screwdriver and re-tighten locknut(C).
4. Don't over-adjust the screw. Over-adjustment will cause shaft in gear box to bend or break.

Height Adjustment.

1. To adjust for height increment indicated in figure 69, pull the spring lever against rubber tire. Rotate wheel and spring lever, together, to the desired slot.
2. If lower cutting height is required, wheel removal is necessary.
3. Remove screw(55-856) and washer(3507), figure 70.
4. Remove wheel, also nut(55-1564).
5. Remove entire height adjustment segment as a unit. Replace in reverse order.

Cleaning the Mower.

1. The chassis is equipped with a water wash hole which enables cleaning underside of mower without turning it over.
2. Clean immediately after mowing.
3. Start engine and run at about 3/4 throttle, out of drive.
4. Insert end of hose into opening with engine running.

Figure 68

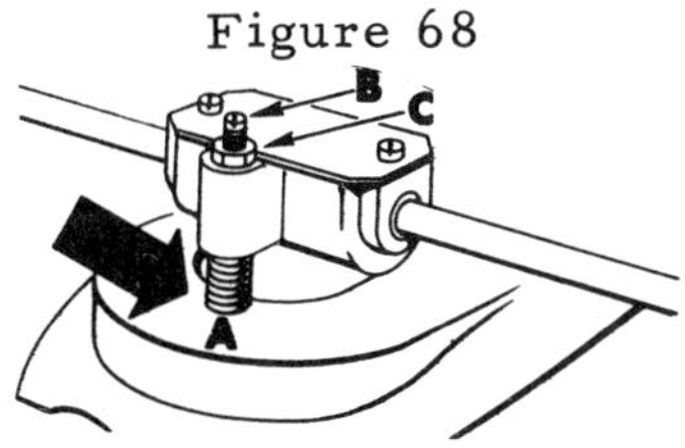

Figure 69

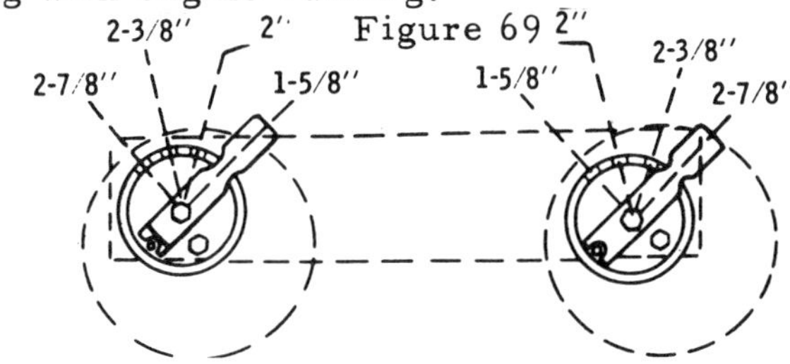

Figure 70
Parts breakdown for Series 8, and 28 Pincor Rotary Lawn Mowers.

MODEL 28S, 28V, 28BH

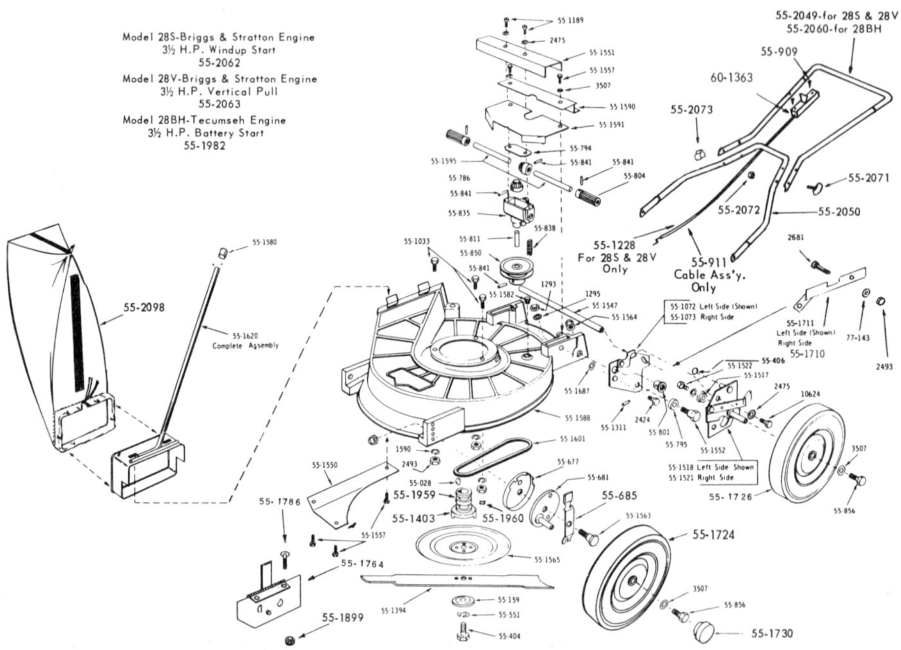

MODEL 8S, 8V

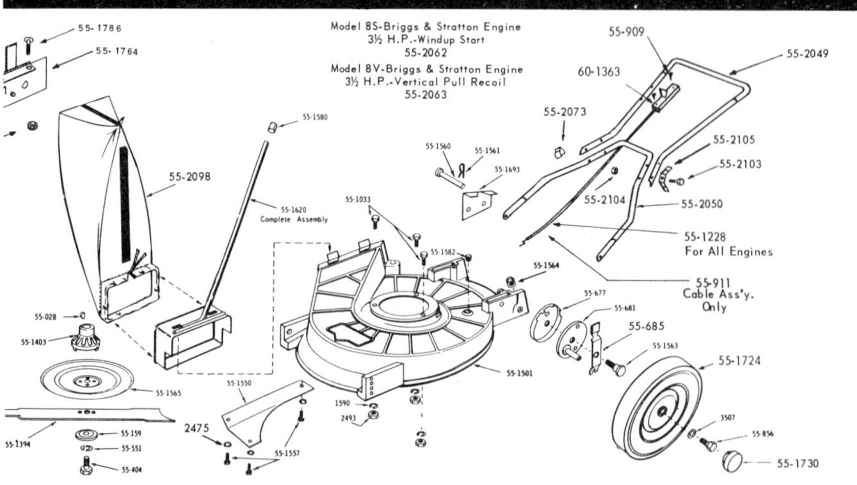

Components and nomenclature for models illustrated on opposite page.

COMPLETE PARTS LIST – MODEL 28S, 28V, 28BH

Part No.	Name of Part	Part No.	Name of Part
1293	Hex Nut	55-1522	Screw
1295	Lockwasher	55-1547	Tie Bar
1590	Lockwasher	55-1550	Chute Strut
2424	Screw	55-1551	Gear Box Shield
2475	Washer	55-1552	Axle Stud
2493	Hex Nut	55-1557	Taptite Screw
2681	Hex Bolt	55-1563	Stud Axle
3507	Washer	55-1564	Cap Nut
10624	Hex Screw	55-1565	Disç
55-028	Woodruff Key	55-1580	Cap
55-159	Washer	55-1582	Cap
55-404	Screw	55-1588	Chassis (Specify Color)
55-406	Washer	55-1590	Belt Guard
55-551	Lockwasher	55-1591	Pully Cover
55-677	Index Plate	55-1594	Sleeve Bearing
55-681	Axle Assembly	55-1595	Drive Shaft
55-685	Position Lever	55-1601	Belt
55-786	Bevel Gear	55-1620	Bag Holder Assembly Complete
55-794	Cover	55-1687	Spacer
55-795	Spacer	55-1710	Handle Bracket, Right Side
55-801	Ball Bearing	55-1711	Handle Bracket, Left Side
55-804	Driver	55-1724	Wheel 8 x 1¾
55-811	Gear Shaft	55-1726	Wheel 8 x 1¾ Plain
55-835	Bear Box Assembly (Includes 2, 55-799 & 2 55-1594)	55-1730	Hub Cap
		55-1764	Chute Safety Plate
55-838	Spring	55-1786	Truss Head Screw
55-841	Roll Pin	55-1899	Hex Nut
55-850	Pully and Hub Assembly	55-1959	Adapter Pully Assembly
55-856	Screw	55-1960	Set Screw
55-909	Sheet Metal Screws	55-1982	Tecumseh Engine 3½ H.P. Battery Start
55-911	Cable Assembly Only	55-2049	Handle, Upper for Models 28S, 28V
55-1033	Hex Bolt	55-2050	Handle, Lower
55-1072	Side Plate Assembly - Left Side (shown)	55-2060	Handle, Upper for Model 28BH
55-1073	Side Plate Assembly - Right Side	55-2062	Briggs & Stratton Engine 3½ H.P. Windup
55-1228	Cable Control for All Engines	55-2063	Briggs & Stratton Engine 3½ H.P. Vertica
55-1311	Set Screw	55-2071	Handle Bolt
55-1394	Blade	55-2073	Cable Clip
55-1403	Adapter	55-2072	Hex Nut
55-1517	Washer	55-2098	Bag Assembly
55-1518	Height Adjustment L.H. (Rear)	55-2130	Mower Manual
55-1521	Height Adjustment Assy. R.H. (Rear)	60-1363	Knob (Cable)
		77-143	Washer

COMPLETE PARTS LIST – MODEL 8S, 8V

Part No.	Name of Part	Part No.	Name of Part
1590	Lockwasher	55-1563	Stud Axle
2475	Lockwasher	55-1564	Cap Locknut
2493	Hex Nut	55-1565	Disc
3507	Washer	55-1580	Cap
55-028	Woodruff Key	55-1582	Water Plug
55-159	Washer	55-1693	Handle Latch
55-404	Hex Screw	55-1620	Bag Holder Assembly Complete
55-551	Lockwasher	55-1724	Wheel 8 x 1¾ Ribbed Tires
55-677	Index Plate	55-1730	Hub Cap
55-681	Axle Assembly	55-1764	Chute Safety Plate
55-685	Position Lever	55-1786	Truss Head Screw
55-856	Hex Screw	55-1899	Hex Nut
55-909	Sheet Metal Screw	55-1982	Tecumseh Engine 3½ H.P. Battery Start
55-911	Cable Assembly Only	55-2049	Handle Upper
55-1033	Hex Bolt	55-2050	Handle, Lower
55-1228	Cable Control for All Engines	55-2062	Briggs & Stratton Engine 3½ H.P. Windup
55-1394	Blade	55-2063	Briggs & Stratton Engine 3½ Vertical Pull
55-1403	Adapter	55-2073	Cable Clip
55-1501	Chassis	55-2098	Bag
55-1550	Chute Strut	55-2103	Handle Bolt
55-1557	Taptite Screw	55-2104	Hex Nut
55-1560	Handle Pin	55-2105	Wing Nut
55-1561	Cotter Pin	55-2130	Mower Manual
		60-1363	Knob, (Cable)

Figure 71
Parts breakdown for Model 4CV self-propelled rotary lawn mower.

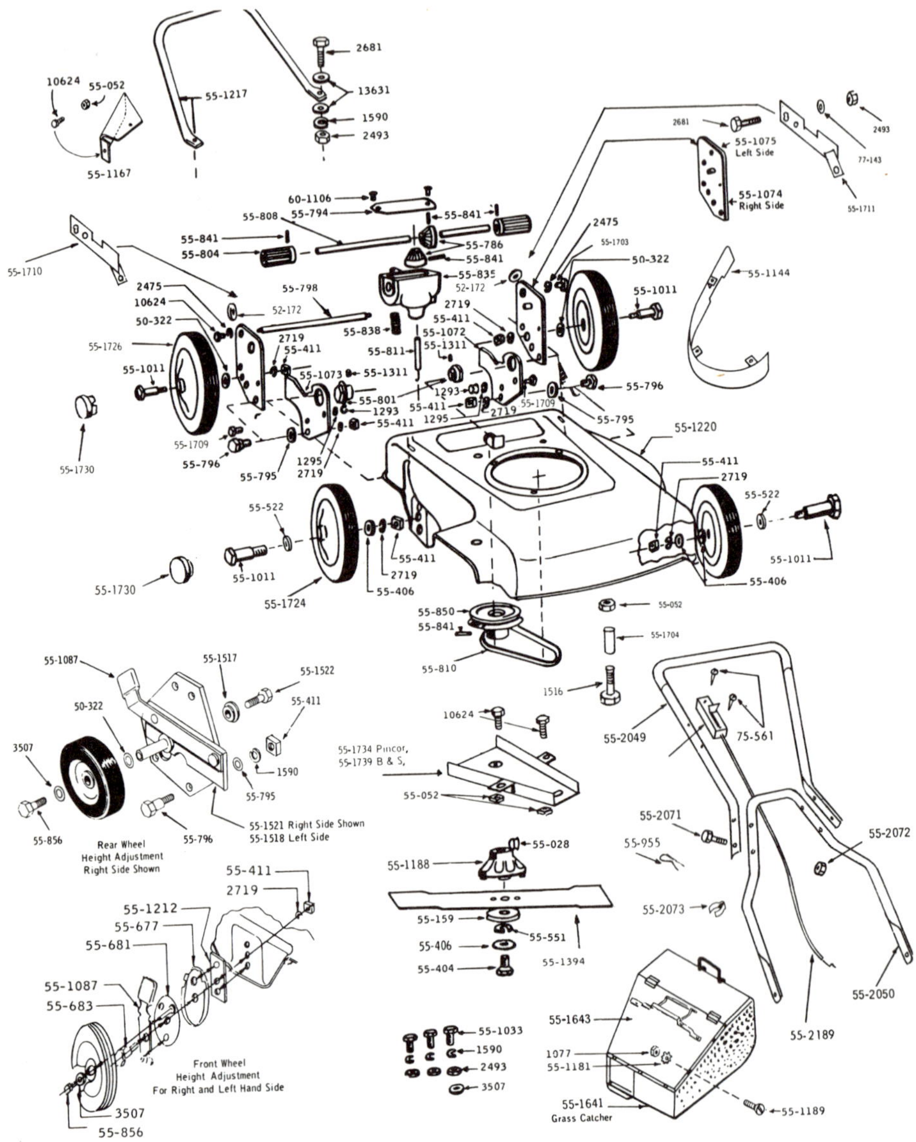

Chapter Four

King O' Lawn by King O' Lawn Inc. Southgate, California.

Mower Operation.

As stated in previous chapters, and is true of all machinery, always read the service manuals covering mower and engine before attempting to use the unit. The King O' Lawn models covered in this chapter utilize the automotive-type differential, as well as a dual clutch system which allows the operator more freedom to manuever the mower, with or without drive clutches in gear. The clutch system is designed to allow wheel drive, and/or reel drive clutches to be operated independently, or at the same time.

To engage cutting reel, press reel drive control knob downward. Reel drive clutch knob is located on right, when standing behind the mower.

The self-propelled mechanism is engaged by pressing wheel drive control knob downward with thumb of your left hand. The drive clutch is designed to allow some clutch slippage, which can create variance in clipping rate, or allow manuevering in tight places. Slippage is created by varying pressure applied to wheel drive control knob. When mowing long, straight runs, engage clutch fully by pushing down and locking knob in place.

Figure 72

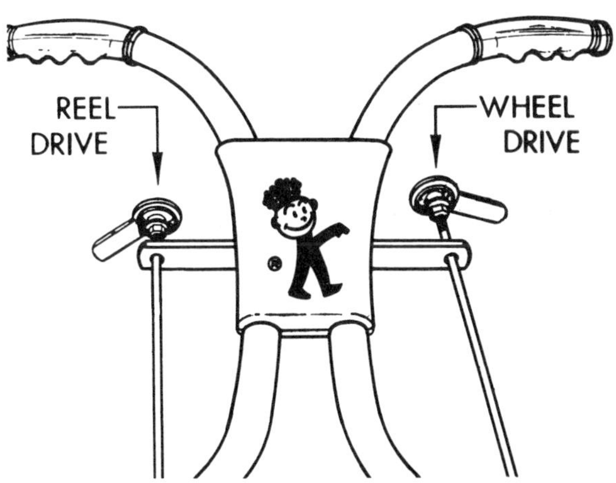

Cutting Height Adjustment.

1. Before making the cutting height adjustment, turn off the engine and move your mower to a level area, preferably a sidewalk or driveway.
2. To make adjustment on low-cut model, refer to figure 73.
3. Position of front wheels determines cutting height. To set, loosen set screw (B) on both wheels. Slide caster assembly (A), up or down to desired cutting height.
4. Retighten set screws(B).
5. To make adjustment on high cut models, refer to figure 74.
6. Cutting height is determined by position of front castor wheels(D).
7. Loosen nuts (E) on both sides of mower and slide bracket (F) up or down to desired cutting height.
8. Retighten nuts(E).
9. When lowering cutting height, always re-check area for foreign objects which may be injurious to your mower and yourself. Remove all stones, sticks, wire, toys etc. before proceeding.

Maintenance.

1. Engine maintenance may be made by turning back to Chapter Two.
2. Refer to figure 75 for mower maintenance.
3. Grease the mower at areas denoted with letter(A).
   a. Castor wheels bodies(low cut models only).
   b. Reel bearings.
   c. Clutch shaft carrier blocks.
   d. Rear axle carrier tube.
   e. Differential housing.
4. Apply grease every 16 operating hours. Oftener in hot climates, or extreme operating conditions such as dusty or humid areas.
5. Apply a lightweight oil(SAE 20) at points(B), every 8 operating hours or oftener under extreme conditions.
   a. Caster wheel and caster swivel bearings in high cut model.
   b. Side plate axle bearings.
   c. Throttle control.
   d. Throttle wire and cable.
   e. All three roller chains.

Figure 73

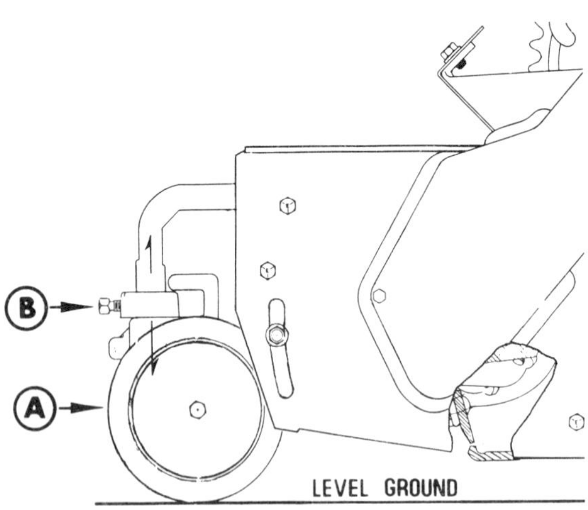

Figure 74

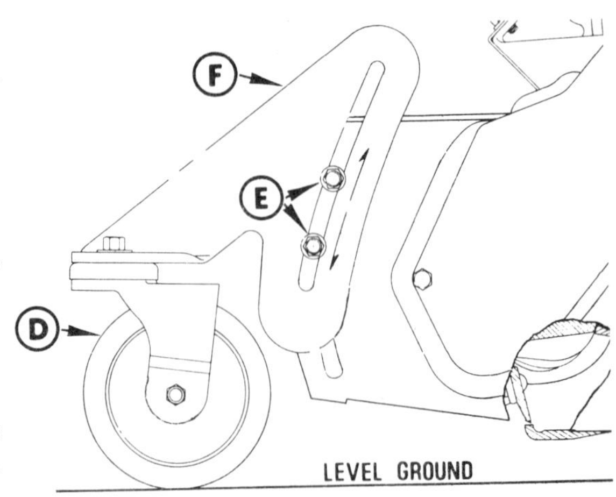

Figure 75

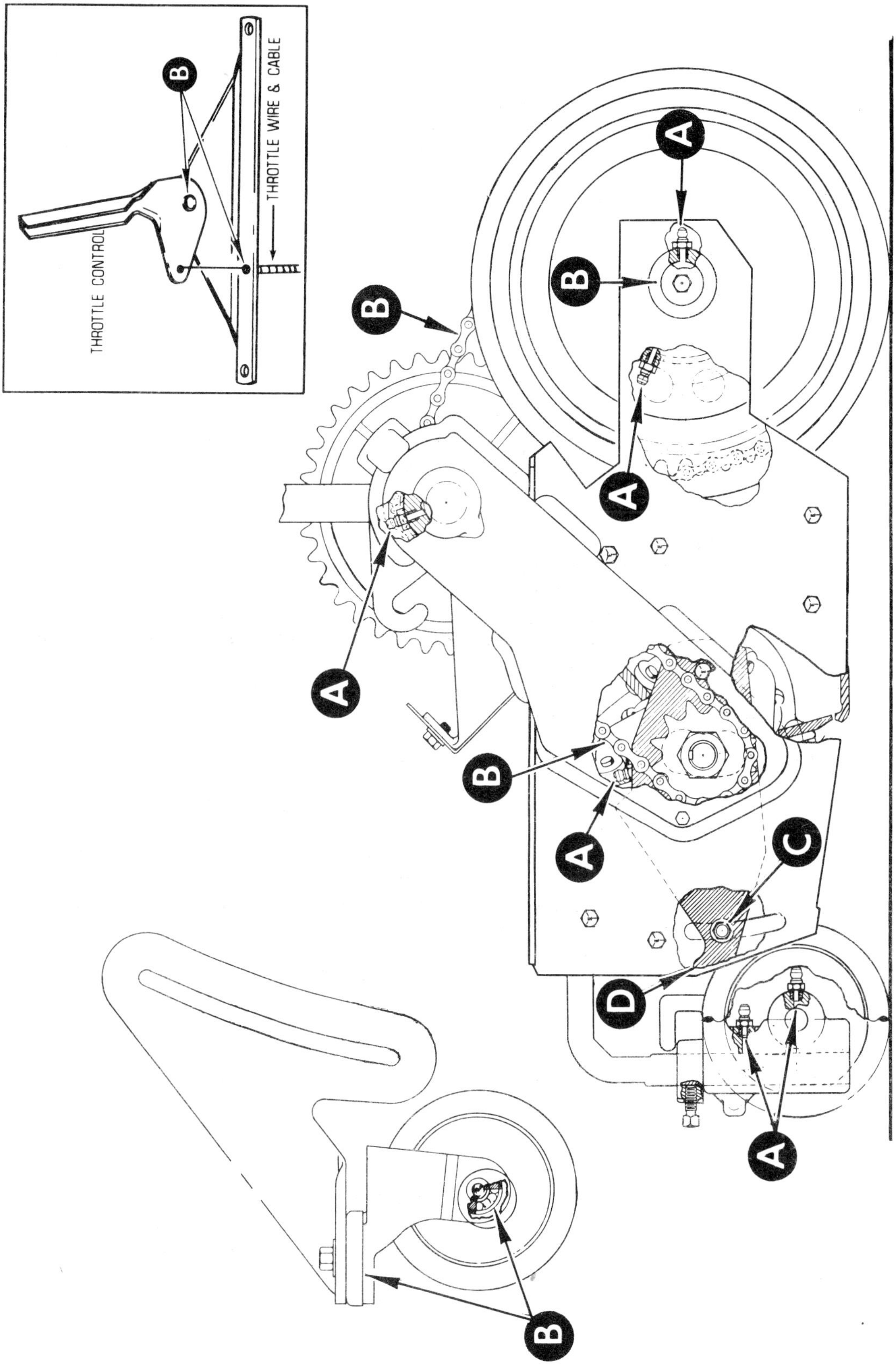

Chain Adjustment.

1. Proper chain care and adjustment are a necessity to prolong working life of the mower.
2. Never overtighten the chain.
3. Don't operate the mower with a too loose, or too tight, chain.
4. Manufacturer states no less than 1/4" flex in chain at midway point between sprockets.
   a. Incorrect adjustment will result in bent shaft, chain and sprocket wear, and cost money to replace.
5. Engine chain adjustment.
   a. Loosen the four engine mounting bolts and slide entire engine in proper direction to correct.
   b. When properly adjusted, tighten one bolt firmly and recheck the flex measurement.
   c. Tighten the other engine mount bolts securely.
6. Reel chain adjustment.
   a. Loosen chain idler located under the chain guard.
   b. Move idler down to tighten chain, and up to loosen chain.
   c. Reel chain must be adjusted every time cutting reel setting is changed.
   d. When chain is properly adjusted, tighten idler.
7. Differential chain adjustment.
   a. Adjustment is rare due to slow travel. If chain becomes loose, link changing is required.
   b. Remove one full link, and add a half link(available at most lawn and garden, or hardware stores). If chain is extremely loose, a full link should be removed.
8. Lubricate chain every 8 operating hours. Oftener in extremely dry climates, or if mower is used rarely.

Cutting Reel to Bed Knife Adjustment.

1. Properly adjusted reel should rotate easily with slight finger pressure. DO NOT OVER-ADJUST.
2. Loosen cap screw(C) and tap downward on reel bearing holder arm at point(D), figure 75, with a small hammer.
3. Push down until reel makes contact with bed knife, yet rotates easily.
4. Adjust the opposite side in the same manner.
5. Retighten cap screw(C).
6. Over-adjustment will cause a rippling of the bed knife.

Replacement of Controls.

1. If replacement of throttle control, or clutch rods, is necessary, refer to figure 76.
2. Install clutch rods by inserting threaded ends up through holes in the throttle control and clutch rod holder.
3. Connect opposite ends to proper clutch cams, right or left hand.
4. Slip cotter pins in holes and bend back to secure.
5. After rods are in place on control console, install control knobs by screwing in place, after a 5/16" hex nut and 5/16" lockwasher have been installed.
6. Turn knob to proper position and hold.

7. Tighten hex nut upward to hold knob in position.
8. Work both controls to be certain clutch assemblies are disengaged when controls are pulled back. If not, adjust as needed.
9. Never start the engine unless both clutches are in OFF position.
10. Throttle cable replacements are similiar on most units.
11. Remove from throttle control, and carburetor. If cable is broken, replace it. If cable is bent, straighten as required and replace.
12. When throttle is in OFF position, setting should correspond on the carburetor.
    a. Refer to engine data in chapter two.
13. After cable is in place, attach cable clamps and snap over handle tube.
14. Check the handle, it must be perfectly straight.
15. To adjust, loosen cap screws, move into position, and tighten the screws.
16. Lubricate components per previous instructions.

Figure 76
Complete handle assembly with clutch and throttle controls.

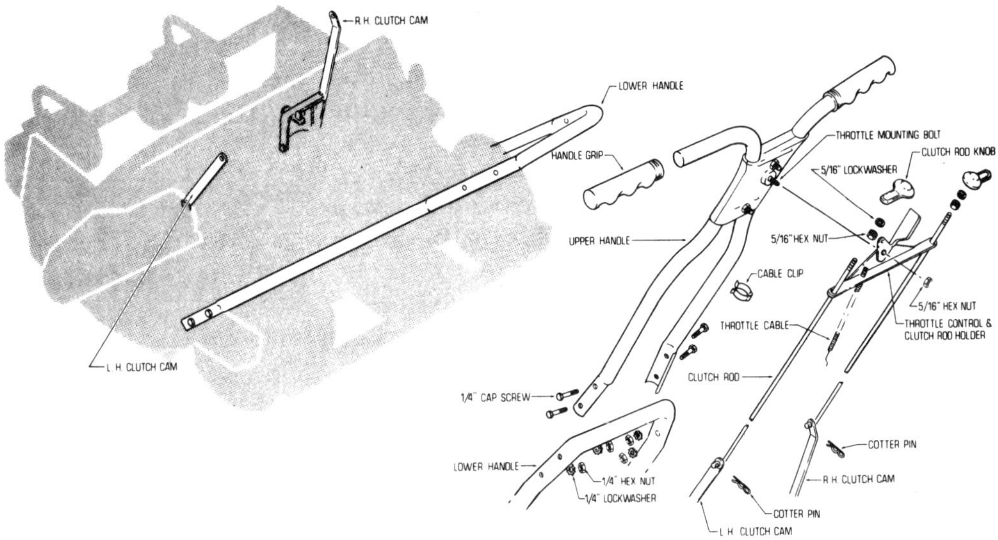

Clutch Adjustment.

The mower is equipped with two independent clutch systems. They are both on the same shaft but must be adjusted individually. Figure 77, illustrates assemblies from front of mower. Reel drive clutch is on the left and wheel drive clutch is on the right.

Adjustment procedures are the same on both clutches, but they MUST be adjusted one at a time.

Reel Drive Clutch Adjustment.

1. Disengage reel drive clutch lever(A).
2. Engage wheel drive clutch lever (B) by pulling lever toward front of the mower.
3. Loosen cap screw (C) but don't remove it. Screw is located under the reel clutch bearing block.
4. Remove locking pin(D).
5. Turn adjusting nut (E) clockwise to tighten clutch, counterclockwise to loosen.
6. Proper clearance between clutch lining and reel drive pressure plate is .020'' to .030''.
7. Replace the locking pin(D).
8. Engage reel drive clutch lever(A) by pulling toward front of mower.
9. If adjustment is correct, tighten cap screw(C).

Wheel Drive Clutch Adjustment.

1. Disengage wheel drive clutch lever(B).
2. Engage reel drive clutch lever(A) by pulling toward front of mower.
3. Loosen cap screw(C) located under reel clutch bearing block, but do not remove.
4. Remove locking pin(F).
5. Turn adjusting nut(G) clockwise to tighten clutch, counterclockwise, to loosen.
6. Adjustment is correct when there is about .020'' to .030'' clearance between lining and wheel drive pressure plate.
7. Replace locking pin (F) and engage wheel drive clutch lever (B), by pulling lever toward front of mower.
8. Tighten cap screw(C).

Figure 77

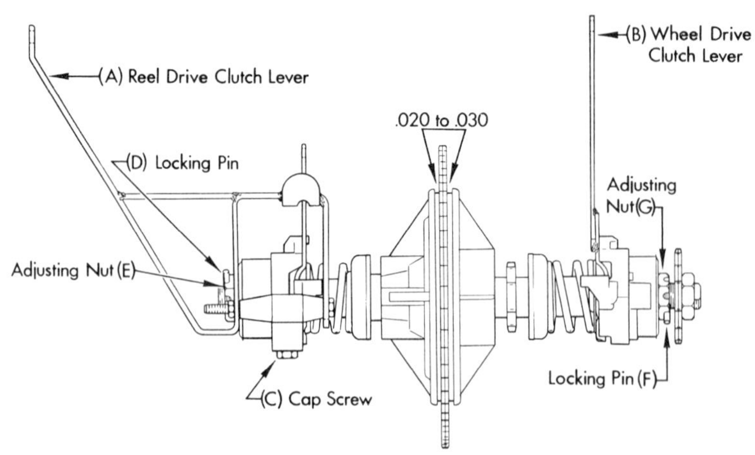

Figure 78
Components breakdown for the clutch assembly.

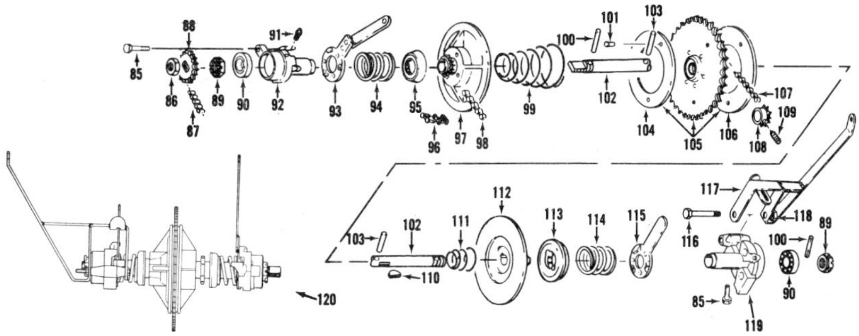

CLUTCH

Key No.	Part No.	Description	425 -5	426 -5	475 -5	476 -5
85	6194	Cap Screw, 5/16" x 3/4"	2	2	2	2
86	2204	Hex Jam Nut, 5/8" sae	1	1	1	1
87	2344	Reel Chain	1	1	1	1
88	7048	Sprocket, 12th (pair)	1	1	1	1
89	7592	Slotted Hex Nut, Special	2	2	2	2
90	7192	Ball Bearing	2	2	2	2
91	1055	Grease Fitting	2	2	2	2
92	7681	Clutch Block, L. H. - 18"	1	1		
92	7682	Clutch Block, L. H. - 21"			1	1
93	7639	Clutch Cam, L. H.	1	1	1	1
94	7462	Pressure Spring, L. H.	1	1	1	1
95	7130	Thrust Bearing, L. H.	1	1	1	1
96	2346	Connecting Link	3	3	3	3
97	7683	Drive Clutch Plate	1	1	1	1
98	7695	Drive Chain	1	1	1	1
99	7133	Clutch Inner Spring, L. H.	1	1	1	1
100	7591	Drive Pin	2	2	2	2
101	7136	Rivets (set of six)	1	1	1	1
102	7575	Clutch Shaft, 18"	1	1		
102	7576	Clutch Shaft, 21"			1	1

Key No.	Part No.	Description	425 -5	426 -5	475 -5	476 -5
103	6228	Drive Pin	2	2	2	2
104	2199	Clutch Linning, L. H.	1	1	1	1
105	7471	Center Sprocket w/linning	1	1	1	1
106	2399	Clutch Linning, R. H.	1	1	1	1
107	2345	Engine Chain	1	1	1	1
108	7395	Sprocket, 8th x 5/8" bore	1	1	1	1
109	1038	Set Screw, 5/16" x 5/16"	1	1	1	1
110	2056	Woodruff Key	1	1	1	1
111	7146	Clutch Inner Spring, R. H.	1	1	1	1
112	7474	Reel Clutch Plate	1	1	1	1
113	7141	Thrust Bearing, R. H.	1	1	1	1
114	7463	Pressure Spring, R. H.	1	1	1	1
115	7494	Clutch Cam, R. H.	1	1	1	1
116	7311	Cap Screw, 5/16" x 3-1/4"	1	1	1	1
117	7690	Bell Crank	1	1	1	1
118	2463	Gripco Lock Nut, 5/16"	1	1	1	1
119	7640	Clutch Block, R. H.	1	1	1	1
120	7691	Clutch Assembly, 18"	1	1		
120	7597	Clutch Assembly, 21"			1	1

Figure 79
Components breakdown for differential assembly

DIFFERENTIAL

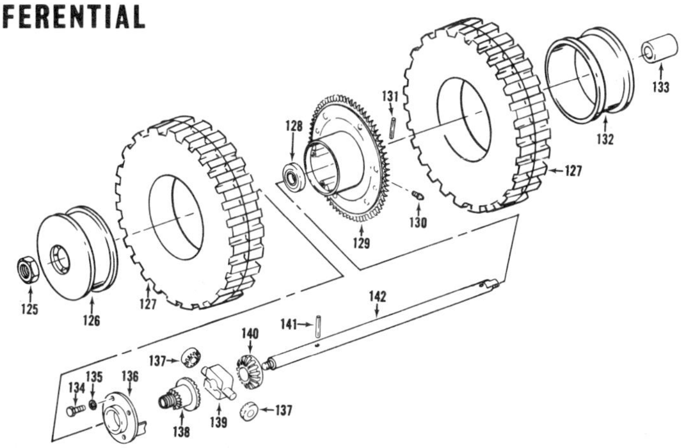

Key No.	Part No.	Description	425 -5	426 -5	475 -5	476 -5
125	6084	Wheel Nut, L. H.	1	1	1	1
126	6075	Rear Wheel, L. H.	1	1	1	1
127	2298	Bar Tread Tire, 10"	2	2	2	2
128	2285	Differential Bearing	1	1	1	1
129	2309	Differential Housing	1	1	1	1
130	1055	Grease Fitting	1	1	1	1
131	7300	Drive Pin	1	1	1	1
132	7089	Rear Wheel, R. H.	1	1	1	1
133	7700	Wheel Spacer, R. H.			1	1
134	1040	Cap Screw, 1/4" x 3/4"	4	4	4	4

Key No.	Part No.	Description	425 -5	426 -5	475 -5	476 -5
135	2226	Lock Washer, 1/4"	4	4	4	4
136	2287	Differential Cap	1	1	1	1
137	2286	Spider Gear (pair)	1	1	1	1
138	6082	Axle & Gear Assy., L. H.	1	1	1	1
139	2283	Cross Shaft	1	1	1	1
140	2304	Differential Gear, R. H.	1	1	1	1
141	2282	Drive Pin	1	1	1	1
142	7629	Differential Axle, 18"	1	1		
142	7630	Differential Axle, 21"			1	1

Figure 80

REEL & BED KNIFE

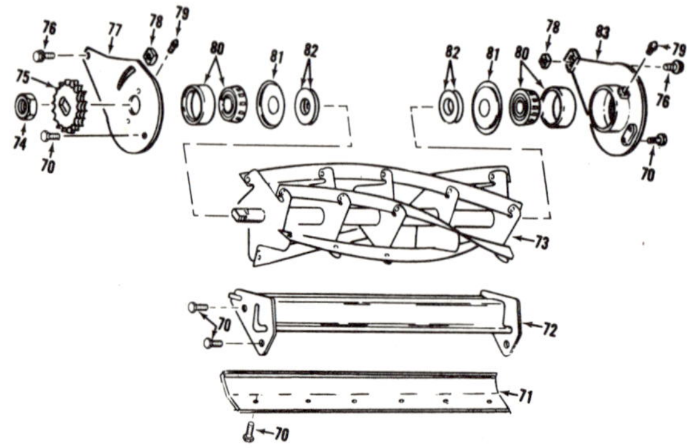

Key No.	Part No.	Description	425 -5	426 -5	475 -5	476 -5
70	1066	Cap Screw, 5/16" x 1/2"	12	12	13	13
71	2413	Bed Knife, 18"	1	1		
71	2417	Bed Knife, 21"			1	1
72	7626	Bed Bar, 18"	1	1		
72	7627	Bed Bar, 21"			1	1
73	7569	Cutting Reel, 18"	1	1		
73	7107	Cutting Reel, 21"			1	1
74	2204	Hex Jam Nut, 5/8" sae	1	1	1	1
75	7658	Sprocket, 9th (pair)	1	1	1	1
76	2410	Cap Screw, 5/16" x 1/2"	2	2	2	2
77	7110	Reel Bearing Holder, L. H.	1	1	1	1
78	2411	Square Nut, 5/16"	2	2	2	2
79	1055	Grease Fitting	2	2	2	2
80	6085	Reel Bearing & Race	2	2	2	2
81	7331	Dust Shield	2	2	2	2
82	7694	Spring Washer	4	4	4	4
83	7119	Reel Bearing Holder, R. H.	1	1	1	1

Figure 81

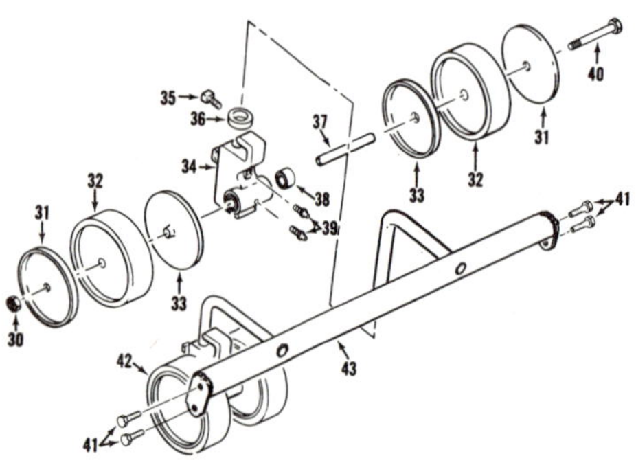

LOW-CUT CASTOR

Models 425-5 & 475-5 Only

Key No.	Part No.	Description	425 -5	475 -5
30	2463	Gripco Lock Nut, 5/16"	2	2
31	2466	Castor Disc, Outer	4	4
32	2610	Castor Wheel	4	4
33	2465	Castor Disc, Inner	4	4
34	2457	Castor Body	2	2
35	1044	Set Screw, 5/16" x 1/2"	2	2
36	7229	Adjusting Collar	2	2
37	7308	Axle Spacer	2	2
38	2461	Ball Bearing	4	4
39	1055	Grease Fitting	4	4
40	7311	Cap Screw, 5/16" x 3-1/4"	2	2
41	1066	Cap Screw, 5/16" x 1/2"	4	4
42	2500	Castor Assembly	2	2
43	6050	Shrub Bar, 18"	1	
43	6051	Shrub Bar, 21"		1

Figure 82

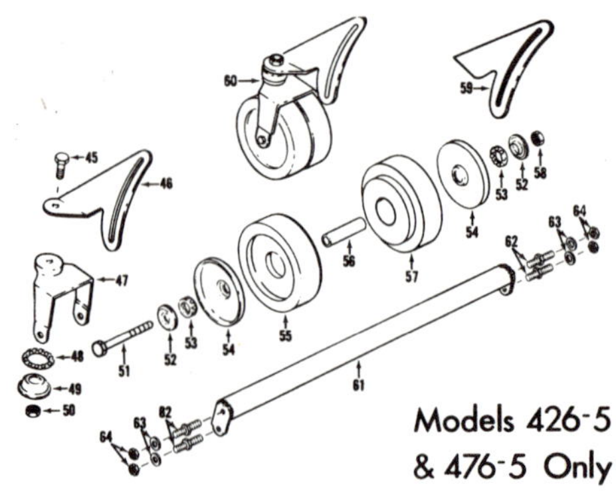

Models 426-5 & 476-5 Only

HIGH-CUT CASTOR

Key No.	Part No.	Description	426 -5	476 -5
45	2624	Cap Screw, 7/16" x 1" sae	2	2
46	7156	Castor Bracket, L. H.	1	1
47	6104	Castor Fork	2	2
48	7147	Swivel Bearing	2	2
49	7157	Race, Swivel Bearing	2	2
50	2625	Hex Nut, 7/16" sae	2	2
51	2818	Cap Screw, 5/16" x 2-3/4"	2	2
52	7158	Race, Wheel Bearing	4	4
53	6168	Wheel Bearing	4	4
54	6129	Castor Disc	4	4
55	2610	Castor Wheel	2	2
56	7160	Spacer	2	2
57	7159	Castor Wheel w/hub	2	2
58	2463	Hex Lock Nut, 5/16"	2	2
59	7161	Castor Bracket, R. H.	1	1
60	7173	Castor Assembly, L. H.	1	1
60	7174	Castor Assembly, R. H.	1	1
61	6053	Shrub Bar, 18"	1	
61	6054	Shrub Bar, 21"		1
62	2475	Double End Bolt, 5/16"	4	4
63	2478	Flat Washer, 5/16"	4	4
64	2019	Hex Nut, 5/16"	4	4

Figure 83

Main Frame assembly.

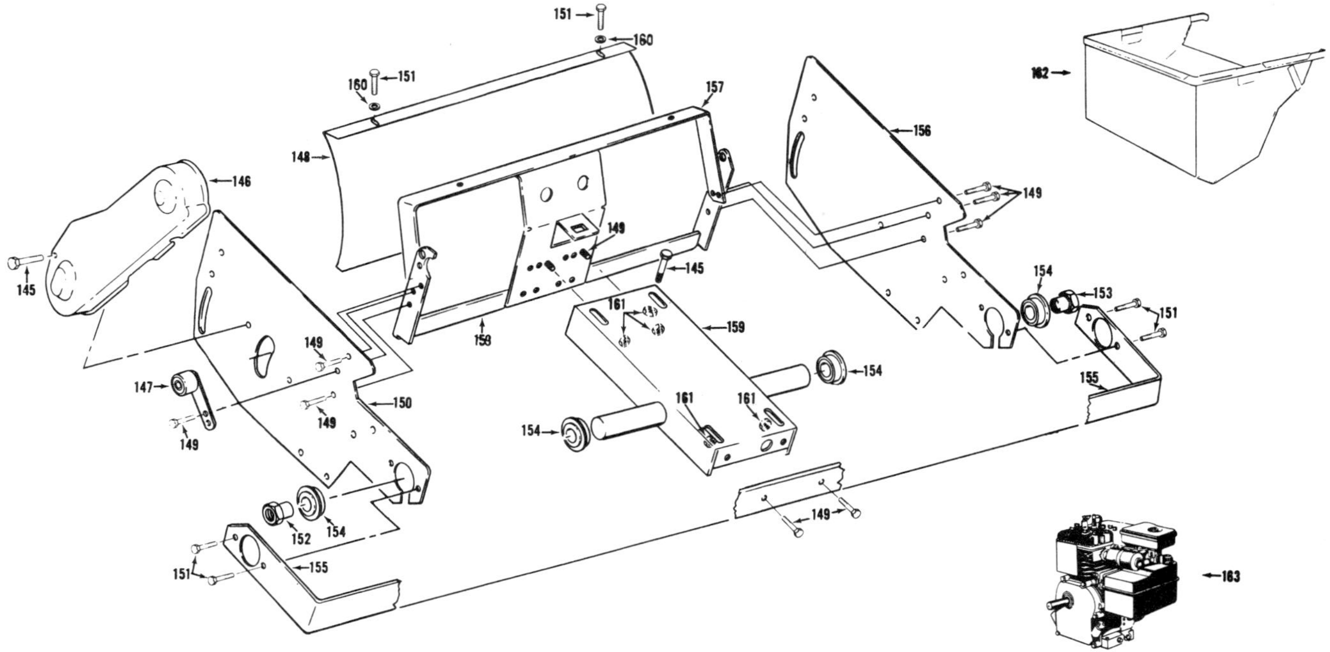

Key No.	Part No.	Description	425 -5	426 -5	475 -5	476 -5
145	1036	Cap Screw, 5/16" x 1-1/2"	5	5	5	5
146	7684	Chain Guard	1	1	1	1
147	2099	Reel Chain Idler	1	1	1	1
148	7708	Grass Shield, 18"	1	1		
148	7709	Grass Shield, 21"			1	1
149	6194	Cap Screw, 5/16" x 3/4"	10	10	10	10
150	7685	Side Plate, L. H.	1	1	1	1
151	1066	Cap Screw, 5/16" x 1/2"	6	6	6	6
152	7653	Rear Axle Nut, L. H.	1	1	1	1
153	7654	Rear Axle Nut, R. H.	1	1	1	1
154	2229	Axle Bearing	4	4	4	4
155	7687	Frame Support Strap, 18"	1	1		
155	7688	Frame Support Strap, 21"			1	1
156	7686	Side Plate, R. H.	1	1	1	1

Key No.	Part No.	Description	425 -5	426 -5	475 -5	476 -5
157	7705	Frame Upper Rail and Clutch Bracket, 18"	1	1		
157	7706	Frame Upper Rail and Clutch Bracket, 21"			1	1
158	7614	Frame Lower Rail, 18"	1	1		
158	7617	Frame Lower Rail, 21"			1	1
159	7701	Motor Plate & Axle Tube, 18"	1	1		
159	7702	Motor Plate & Axle Tube, 21"			1	1
160	2478	Flat Washer, 5/16"	2	2	2	2
161	2019	Hex Nut, 5/16"	8	8	8	8
162	7697	Grass Catcher, 18"	1	1		
162	7698	Grass Catcher, 21"			1	1
163	7199	B & S 3hp Engine	1	1	1	1

Model 246-3, King O' Lawn Edger.

Operating controls.

1. Blade clutch(A) disengages cutting blade when pressed down with foot. To engage cutting blade, flip lever up with your foot.
   a. Always disengage cutting blade when starting engine.
2. Throttle control (B) determines speed. Stop position is at extreme left. Choke position is at extreme right. Intermediate position of the lever determines operating speed.
3. Cutting Depth Control (C) determines depth of cut. Pull lever back for deeper cutting depth. Push lever forward for less.
4. Cutting Blade Angle Control (D) changes cutting angle. To adjust, loosen wing nut, and push lever forward, or pull lever back to desired cutting angle and retighten wing nut.
   a. Illustration shows cutting blade in a vertical position, but it can be adjusted from a reverse angle to a complete horizontal position. Adjustments may be made with engine running or off, but from safety standpoint, OFF is safer.
   b. Cutting depth adjustment may be necessary after angle adjustment has been made.

Maintenance.

1. Engine maintenance procedure may be found in Chapter Two.
2. Grease should be applied to areas(1 & 2), every 8 hours of operation.
3. Oil, use SAE 20, all pivotal points, wheel hubs and throttle control-cable.

Changing the Cutting Head.

1. Set cutting head in vertical position.
2. Turn blade until it's parallel with the ground.
3. Lower depth control until cutting head rests on the ground.
4. Push down on cutting blade guard and hold while loosening the nut.
5. Replace changed cutting head and tighten holding nut.

Figure 84

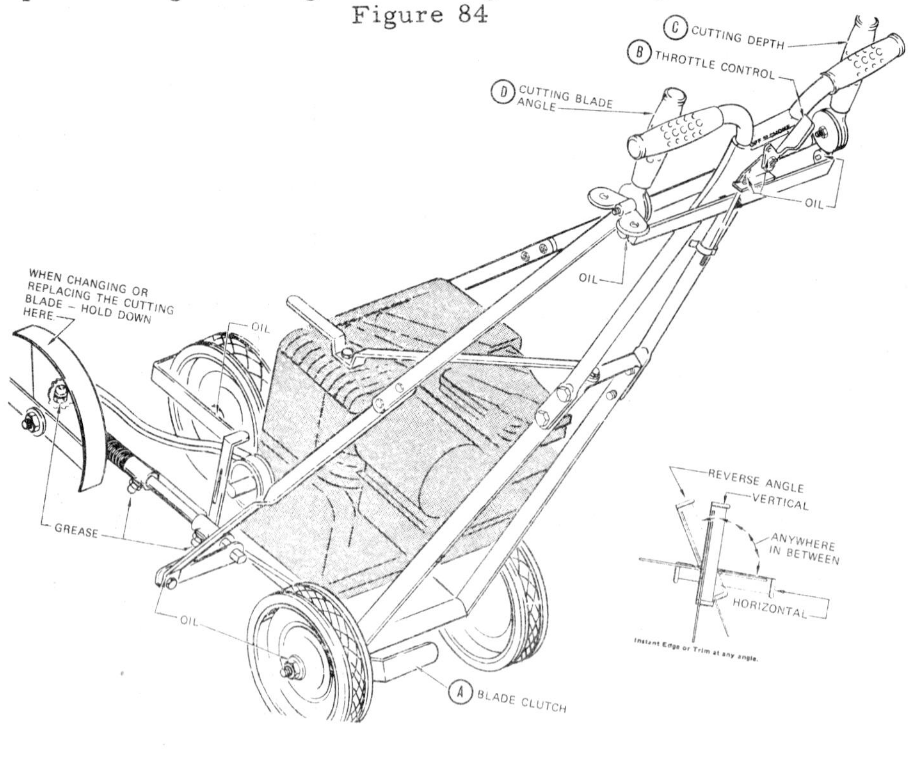

Figure 85
Cutting head components and parts listing.

CUTTING HEAD ILLUSTRATION

CUTTING HEAD PARTS LISTING

MODELS 20A-3, 20AD-3 & 246-3

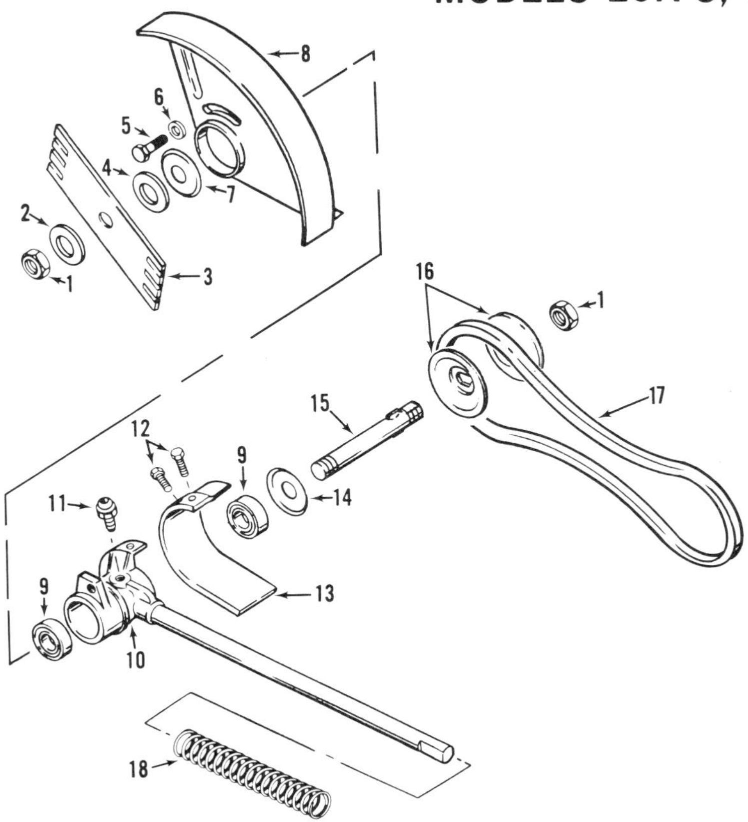

MODELS 20A-3, 20AD-3 and 246-3

KEY NO.	PART NO.	DESCRIPTION	QUAN. USED ALL MODELS
1	1018	Hex Jam Nut, 1/2" sae	2
2	1017	Blade Washer, 1/2" i.d.	1
3	1096	Cutting Blade	1
4	2638	Blade Washer, 1/2" i.d.	1
5	1066	Cap Screw, 5/16" x 1/2" std	1
6	2478	Flat Washer, 5/16" i.d.	1
7	6170	Bearing Shield, 1/2" i.d.	1
8	6163	Blade Guard	1
9	7192	Ball Bearing, Metric	2
10	7673	Head Mandrel	1
11	1055	Grease Fitting	1
12	1094	Cap Screw, 1/4" Self-tap	2
13	1104	Belt Guard	1
14	7020	Bearing Shield, Metric i.d.	1
15	7659	Head Shaft	1
16	6066	Split Pulley (pair)	1
17	1106	"V"-Belt, 27"	1
18	1103	Head Tension Spring	1
	7674	Complete Cutting Head Assembly	1

FRAME & HANDLE ILLUSTRATION

MODELS 20A-3, 20AD-3 & 246-3

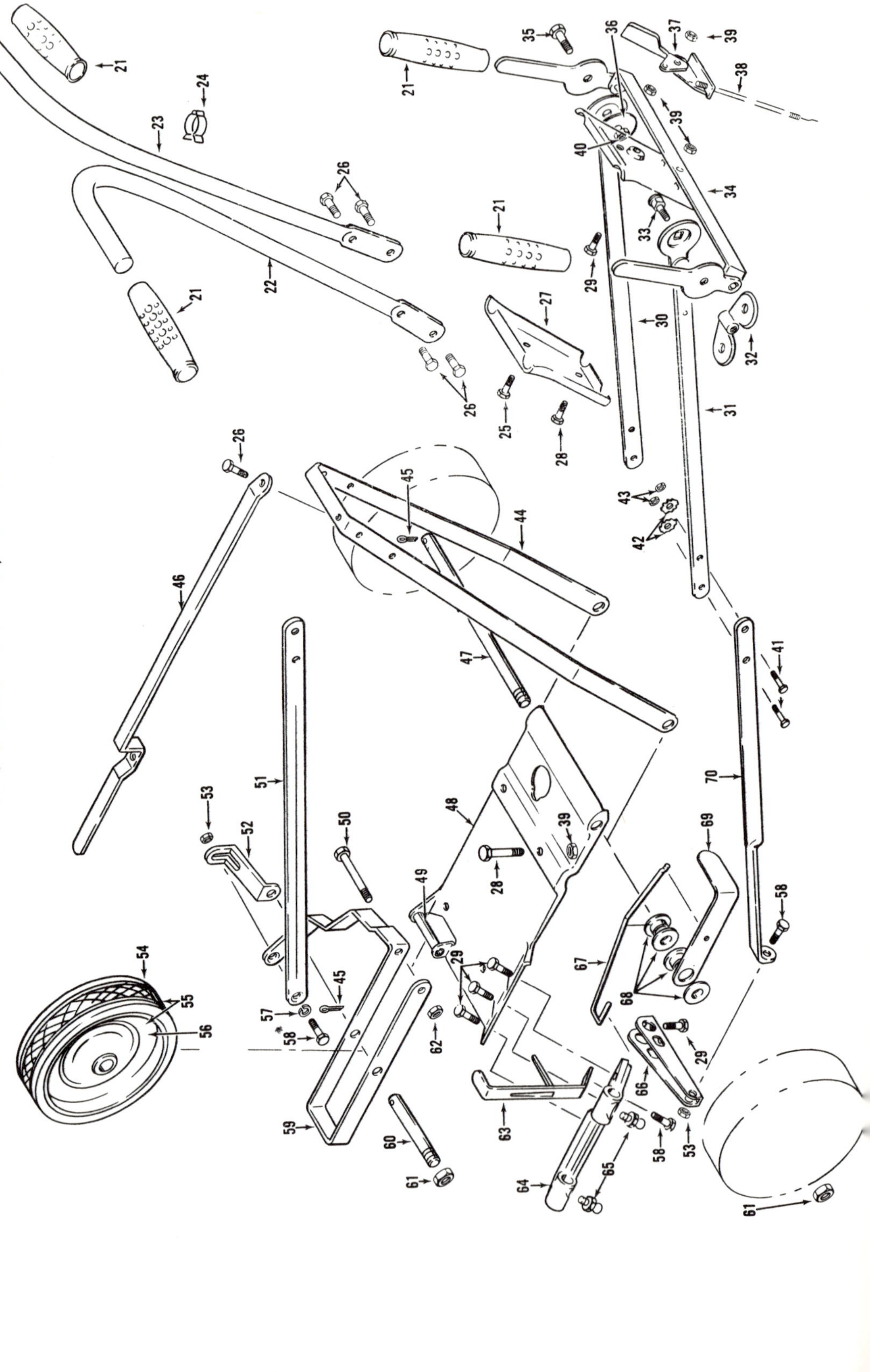

Chapter Five

Montgomery Wards.

Wards market a variety of mowers of various types. One unit covered in this book is a 20" Rotary "Garden Mark", with a Briggs & Stratton engine.

At this stage of the book it should be fairly well established that most units operate basically alike. Service procedure covering engine repair, sharpening blades, and changing cutting heights, are universal. If your particular model isn't covered, check identity for a similiar unit and proceed accordingly.

Operation.

1. Always use clean, fresh regular gasoline. Don't mix with oil.
2. Do not fill gasoline tank while the engine is running.
3. Use a good grade of detergent oil, SAE 30 in warmer climates, and SAE 20 for cooler climates. "For Service MS" is recommended.
4. Fill crankcase per manufacturers instructions.
5. Add oil to air cleaner and clean frequently.
6. The oil should be changed every 25 hours of operation, after initial breaking in period, which should be after first 5 hours of use.

Figure 87

Briggs and Stratton Engine used on several Wards mowers.

7. Pour oil in slowly. Capacity is 1 1/4'' pints.
8. Check dipstick and DO NOT overfill.

Figure 88

Figure 88a

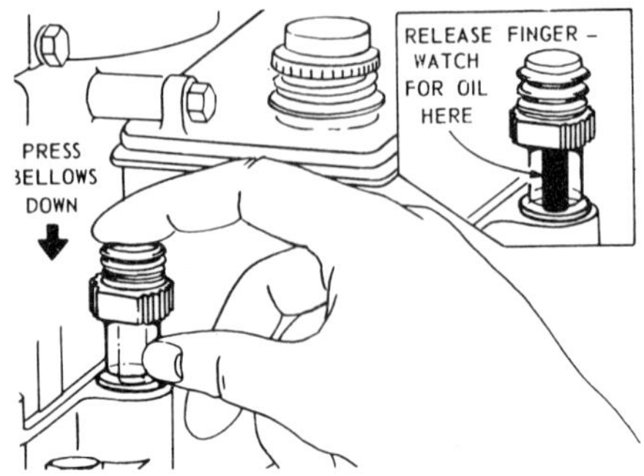

1. Figure 88a illustrates visual oil gauge which allows checking the oil level without removing filler plug.
2. Stop engine and mower must be moved to level area.
3. Press down on the bellows
4. If tube fills with oil, engine is within safe operating range.
5. If tube doesn't fill with oil, add oil until crankcase overflows.
6. To drain oil when change is required, remove oil drain plug on the underside of mower(figure 89).
7. Mowers with engine shown in figure 90 may drain in that manner.

Air Cleaner Service.

1. Service air cleaner at least every 25 operating hours.
2. Remove unit(figure 91), by removing the screw.
3. Clean unit thoroughly and re-oil. Wash element in kerosene or in a petroleum solvent. Squeeze dry and blot until all cleaner is removed.
4. Re-oil element with 3 tablespoons of engine oil. Squeeze element until oil is evenly distributed throughout the element.
5. Re-assemble and attach in place over carburetor. Use caution and don't allow dirt or foreign matter in carburetor during removal, or replacement of element.

Figure 89

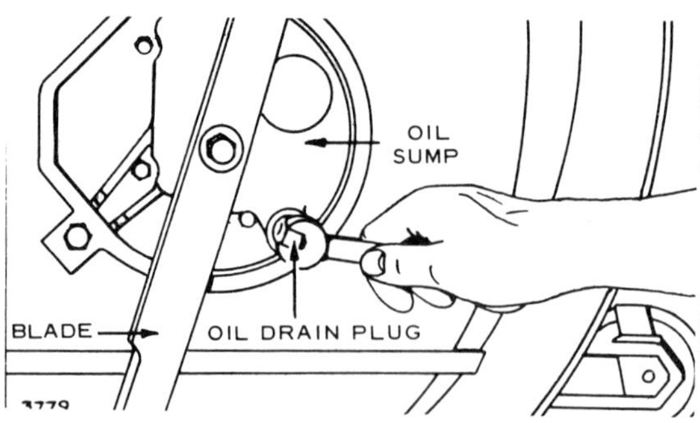

Figure 90

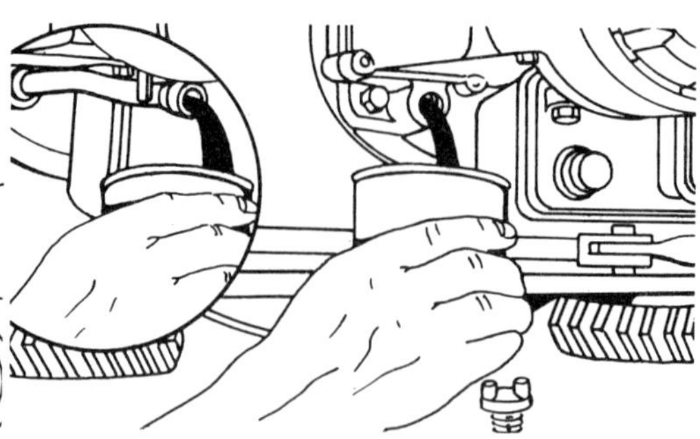

Operation and Carburetor Adjustment.

1. Most units utilize a remote choke-throttle control on the handle.
2. Before starting, check Choke-A-Matic on engine. Remove the air cleaner and push control to CHOKE position. Butterfly in engine carburetor must close entirely(figure 92).
3. Move dial control to START position, if control is in adjustment.
4. If adjustment is required, turn dial back to STOP position.

Figure 91

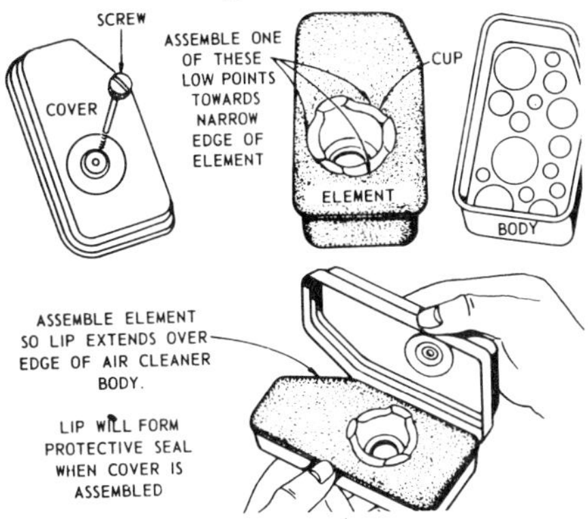

Figure 92

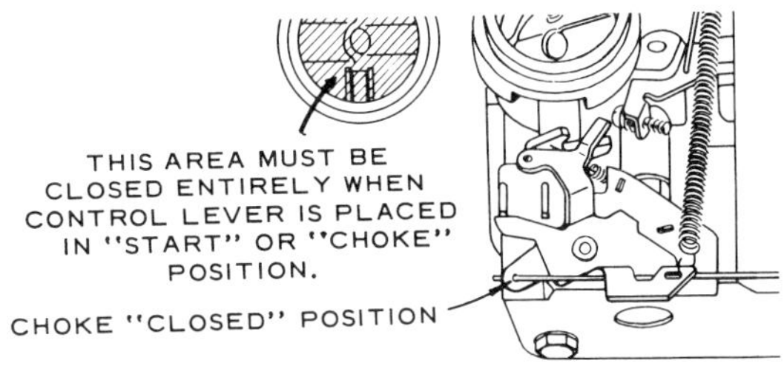

1. Adjustments are primarily for powered mowers.
2. Remove the air cleaner
3. Move remote control lever to CHOKE position, fig. 94.
4. Carburetor should resemble illustration in figure 92.
5. Move control lever to "STOP". Control lever should make contact with stop switch(fig. 94). Choke is overcenter type for positive choke.
6. If not, place remote control lever on handle in FAST position.
7. Lever(C), figure 94, should just touch choke arm at point(D).
8. To adjust, loosen clamp screw(A) on housing.
9. Move control casing(B) proper direction to correct. Tighten screw (A) and recheck controls for proper configuration.
10. Replace the air cleaner.

Figure 93

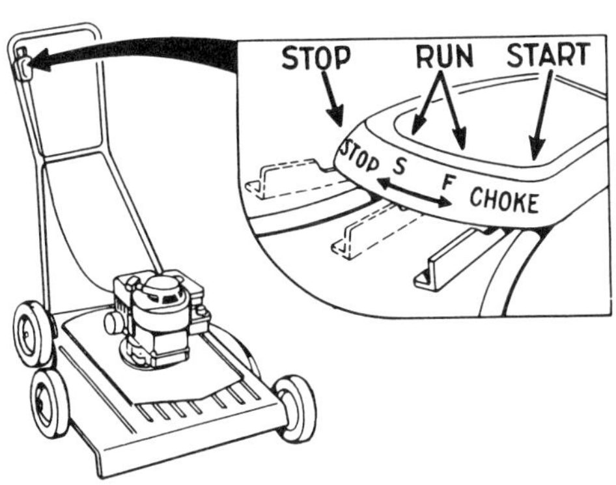

Figure 94

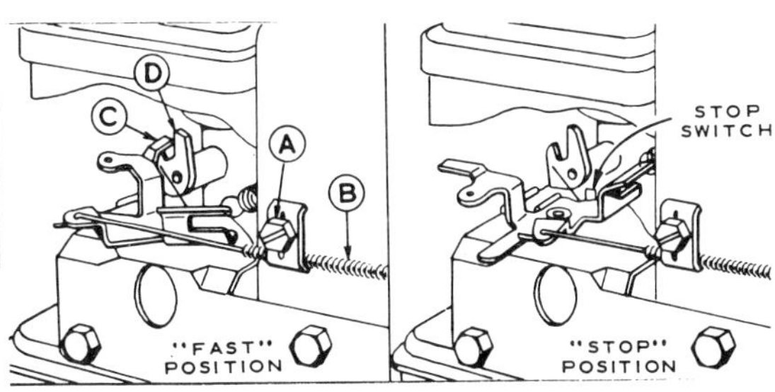

1. Choke-A-Matic controls on housing require adjusting only if housing has been removed.
2. To adjust, place dial control in START position.
3. Loosen control wire screw(A), figure 95, and move lever (C) to full choke position.
4. Allow a slight gap(1/8''), as shown in figure 95.

Figure 95

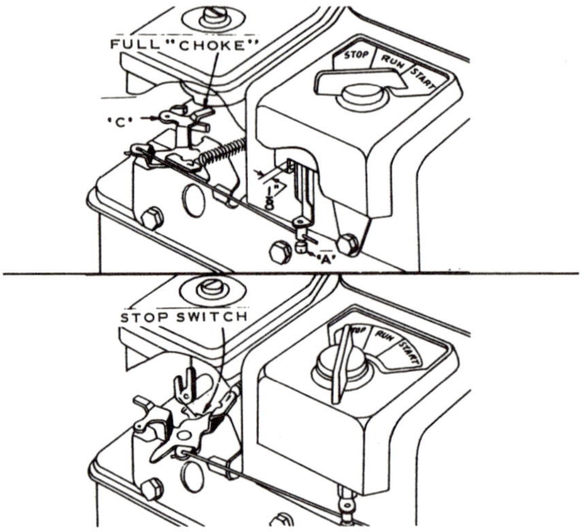

Carburetor Adjustment.

1. Do not remove air cleaner to adjust carburetor.
2. Turn needle valve inward completely(figure 96).
3. Turn back 1-1/2 turns, which is close enough to allow engine to start before making the final adjustment.
4. Start engine and run at approximately 3000 RPM's (normal running speed), with a load.
5. Turn needle valve clockwise until engine starts losing speed, because mixture is getting lean.
6. Turn needle valve counterclockwise until engine runs smoothly, turn past this point until engine again runs unevenly. At that point engine will operate smoothly under load.
7. Hold throttle in idle position and turn idle adjusting screw until idle increases to about 1750 RPM's.
8. Test engine under full load. If it stalls or dies, mixture is still too lean. The richer mixture might result in some uneveness during idle.

Figure 96

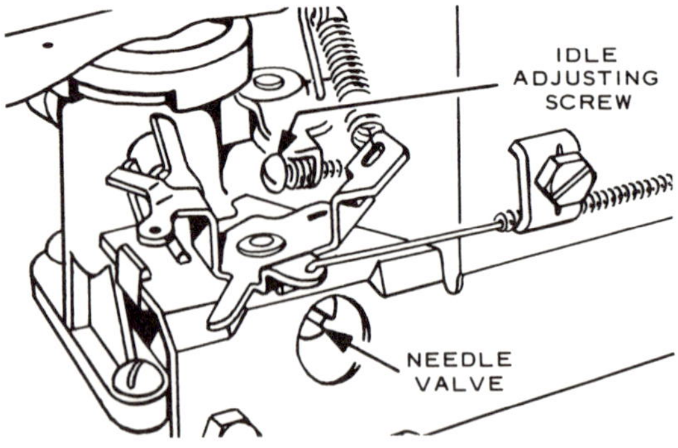

Figure 97

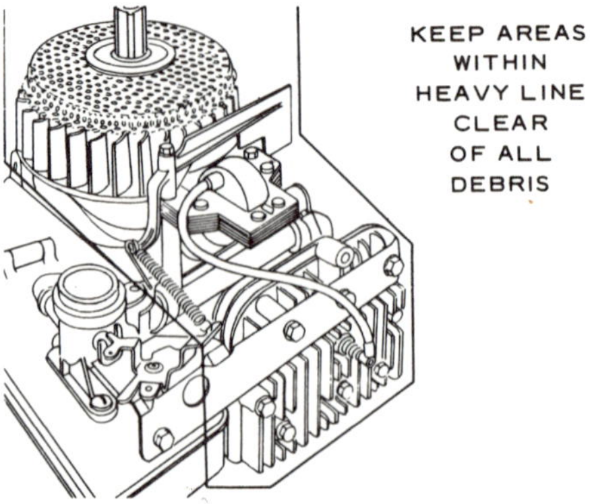

Cleaning the Cooling System.

1. Keep engine clean during operation.
2. Refer to figure 97, and clean all foreign matter away from head, flywheel, and carburetor. Clogged cooling system will result in over-heating and severe damage to the engine.

Specifications for Tune-UP.

1. Spark plug gap should be approximately .030".
2. Ignition point gap should be .020".
3. Intake valve clearance .005" - .007".
4. Exhaust valve clearance .009"- .011".

Starters.

1. Rewind starter is similiar to unit covered in previous chapter on a Pincor mower.
2. Wind-up starter is similiar to unit covered in Pincor chapter.
   a. Never work on mower blades, or under the machine, while the wind-up unit is in wind-up position.

Mower Adjustment.

1. To change cutting height adjustment, remove wheel from frame and move to desired hole. Place each wheel in same hole height to attain uniform cut.
   a. If wheel is placed in lower hole, a higher cutting height will result. Place wheel in higher hole to lower cutting height.
2. If higher handle height is required, reverse the upper handle.
3. To adjust throttle control wire assembly, loosen screw which holds assembly at engine. Move control wire to "CHOKE" position. Move lever, where control wire is attached on engine, to full choke position and retighten the screw.
   a. Oil throttle control and wire, occasionally with light oil.
4. Before cleaning underside of mower, stop engine and dis-connect the spark plug wire.
5. Turn mower on its side and clean all foreign matter away. If you use the water hose, wait until engine is cool and avoid directing water on the engine.
6. To remove blade, turn mower on side.
   a. Remove bolt and washer holding blade to engine crankshaft.
   b. Remove blade and hub assembly.
   c. Remove bolts and washers holding blade to hub assembly.
   d. To sharpen blade, follow original angle of grind. Grind each of the cutting edges equally. Over-grinding will create an imbalance which could be harmful to mower engine and cut unevenly. To check for proper balance, hang on a nail and position in dead center. If blade falls one direction, grind more away from that side. Make grinding in several small steps.
7. Check all nuts and bolts frequently. Keep them tight.

Figure 98

Figure 99

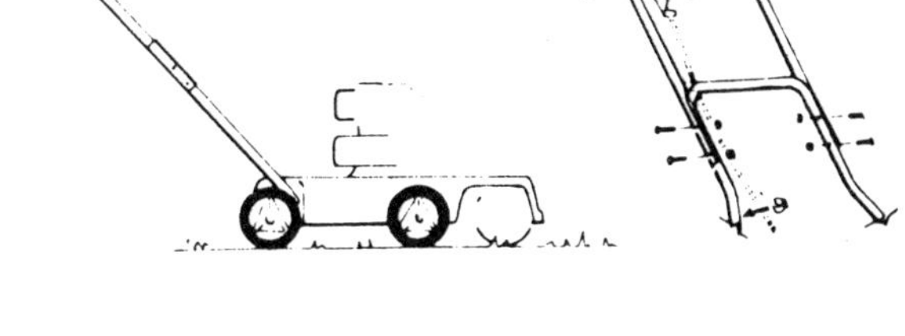

Wards "Garden Mark" Reel Mower-ModelZYJ-197A.

The Model ZYJ-197A is a reel type mower which can be operated as a self-propelled unit, or manual push. With a console-type control panel all controls are within easy reach of the operator.

Engine service and maintenance procedures are similiar to those covered at the beginning of the chapter. A few points of difference will be covered after mower service procedures.

1. Follow all basic steps before starting the engine.
2. To start engine, move control lever to Ease-O-Matic position, with the handle down.
3. Set throttle lever in START position.
4. Pull starter rope quickly once, and move throttle to FAST position.
5. Move throttle lever to "SLOW" when engine starts, and regulate the speed as required.
6. To operate, place operating control lever in Ease-O-Matic position.
7. Raise handle to start, lower it to stop.
8. Place operating control lever in Manual position for direct drive.

Cutting Height Adjustment.

1. Place mower on flat surface such as driveway or sidewalk.
2. Tilt machine back on rollers until wheels are off the ground.
3. Move right-hand wheel lever(B) outward to disengage locating pin. Move lever to desired setting(see figure 102). Repeat on opposite side and be certain both settings are the same.
4. Tilt mower forward until rollers leave the ground. Turn right hand roller knob(A), counterclockwise 2 complete turns and move knob up or down until alignment hole in roller bracket is in line with proper number on alignment chart.
5. Turn knob clockwise until tight. Repeat on the other side.

Ease-O-Matic Adjustment.

1. To adjust Ease-O-Matic, place lever in Manual position.
2. Lower handle to operating position.
3. Loosen adjusting screw locknuts(figure 101a).
4. Push handle down and adjust screw until Actuator Roller has 1/16" clearance between roller and jackshaft.
5. Tighten adjusting screw locknut.

Figure 100

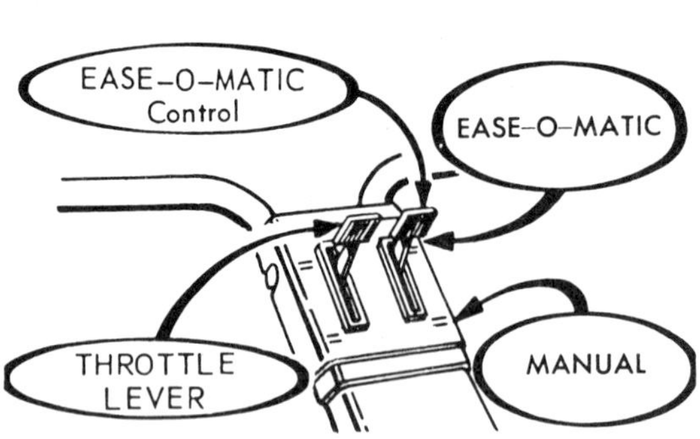

Figure 101

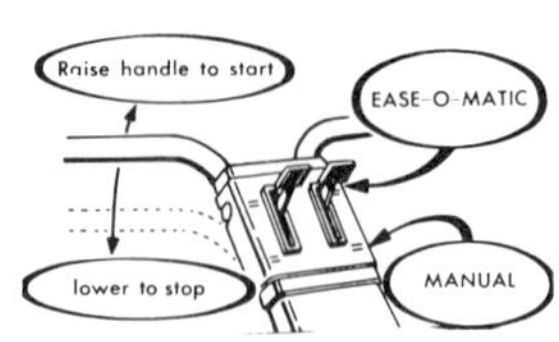

Figure 101a

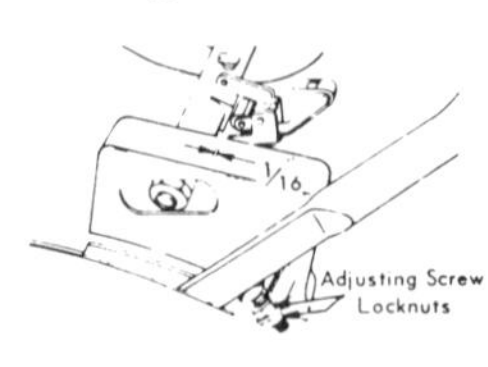

Figure 102

CUTTING HEIGHT IN INCHES	(A)	(B)
2¼	4	5
2	4	4
1¾	3	4
1½	2	4
1¼	2	3
1	1	3
¾	1	2
½	1	1

Figure 103

Cutter Bar Adjustment.

1. Cutter bar should almost contact reel blades when properly adjusted.
2. To adjust, check figure 104, and make adjustment with bolts 2 & 3.
3. To tighten cutter bar(bring closer to reel blades), turn hex adjusting nut counterclockwise as needed.
4. To loosen cutter bar(move away from reel blades), turn hex adjusting nut clockwise as needed.
5. One simple way to check adjustment is the newspaper test. Cut a strip of newspaper 2" wide and 6" long. Insert between cutter bar and reel blades as illustrated in figure 105.
   a. Turn reel backwards slowly. If paper is held by both blades, the adjustment is correct.
   b. If re-adjustment is needed, adjust cutter bar until reel blades pinch the paper without cutting it. Refer to steps 3 & 4 above.
6. To sharpen cutter bar, remove by loosening bolt(3), fig. 104.
   a. Turn left hand bolt counterclockwise, right hand bolt clockwise.
   b. Follow previous steps regarding cutter blade and sharpen.

Figure 104

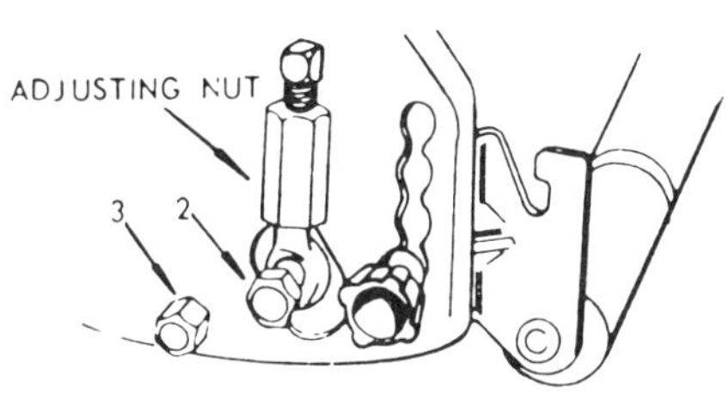

Figure 105

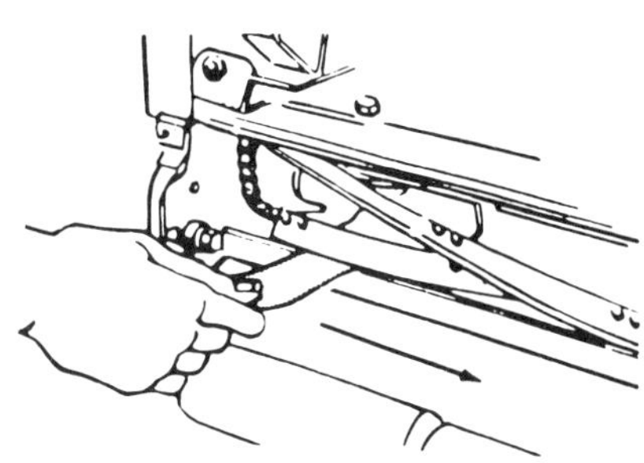

Chain Drive Adjustment.

1. Set hand control in Ease-O-Matic position with handle in standup position.
2. Loosen all four engine mount bolts and pull engine back. Fig. 106.
3. Loosen 4 control assembly bolts and move spacer plate under control assembly.
4. Pull entire assembly back to tighten the chain.
5. When the chain is pinched together, slackness should measure 1/8" from normal tracking of the chain.
6. Tighten three control assembly bolts. Leave spacer plate loose for the moment, until belt is adjusted.

Figure 106

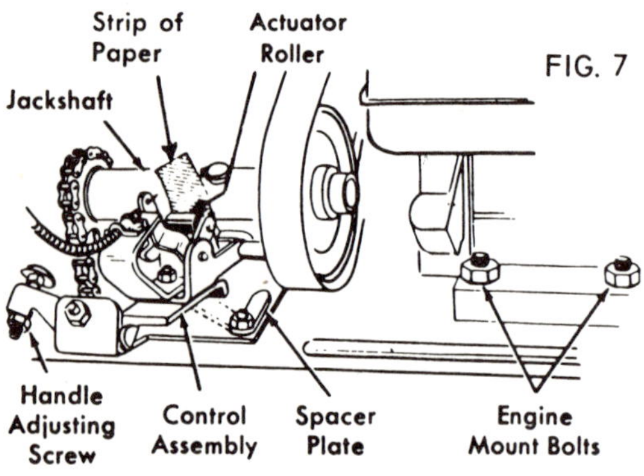

Figure 107

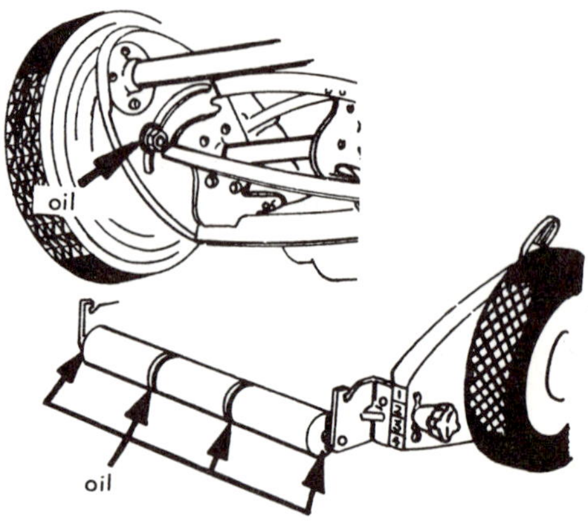

Belt Drive Adjustment.

1. Follow the same steps as in chain drive adjustment.
2. Engine mount bolts must be loose and spacer plate positioned under control assembly(figure 106).
3. Place a slip of paper between jackshaft and actuator roller.
4. Hold control assembly firmly against the spacer plate.
5. Move engine forward until the slip of paper is just barely free.
6. Tighten engine mount bolts.
7. Remove spacer plate.
8. Tighten spacer plate nut.

Lubrication.

1. Engine should be lubricated per previous instructions.
2. Oil jackshaft oiler cup, roller(points illustrated), and roller shaft , with regular motor oil.
3. Oil all friction points on end plate assemblies.
4. Wheel bearings and reel bearings should be lubricated seasonally.

Choke-A-Matic.

1. To adjust, place remote control lever on handle in FAST position.
2. Loosen control casing clamp screw(B) on carburetor, fig. 110.
3. Move control casing(A) and wire proper direction, until speed lever (C) just touches choke operating link at point(D). See figure 110.
4. Tighten casing clamp screw(B).

Figure 108
Parts breakdown on Wards "Garden Mark" -18" Reel-type Lawn Mower.

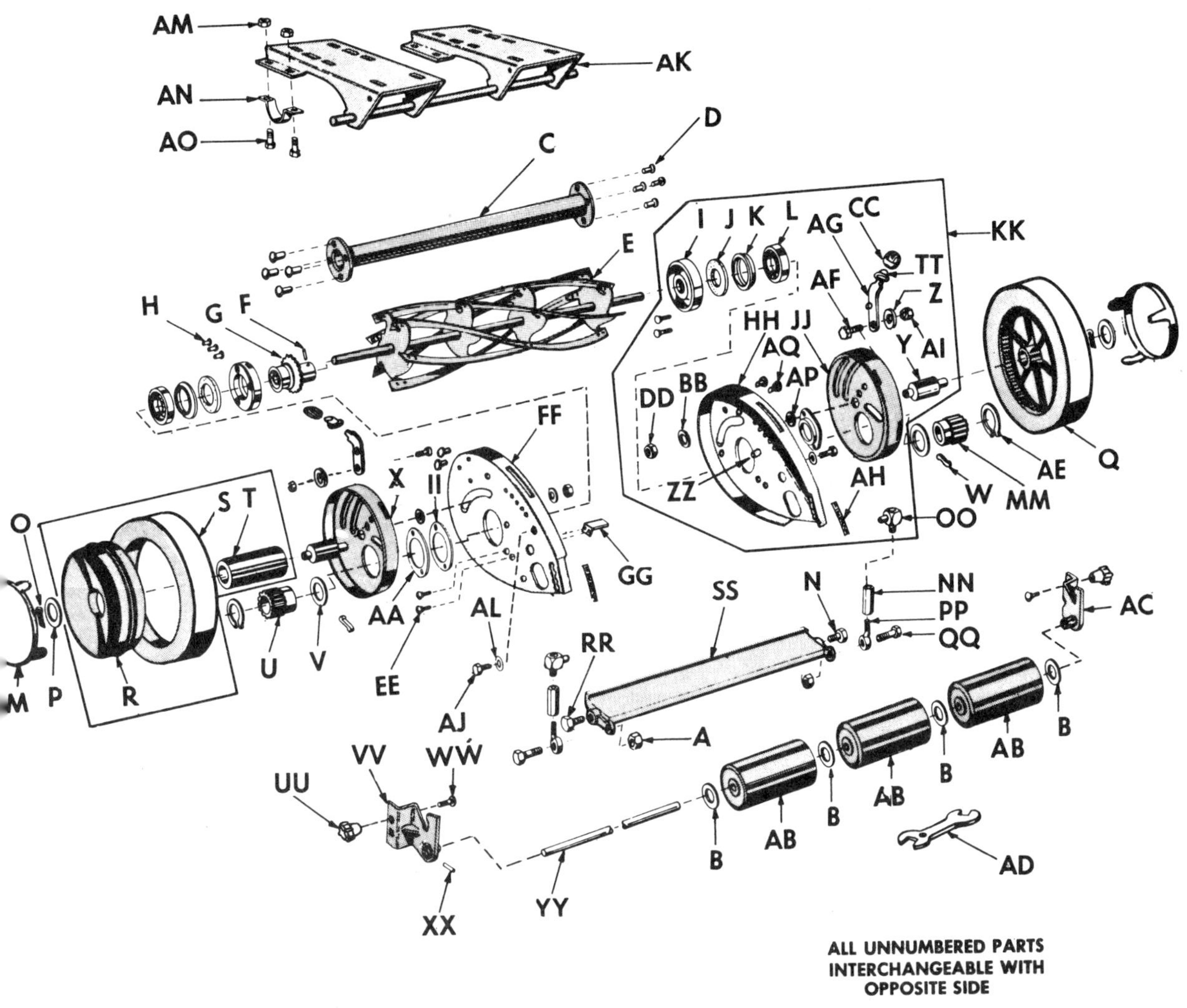

Figure 109
Frame, Handle, Engine and Drive Components- "Garden Mark"- Wards

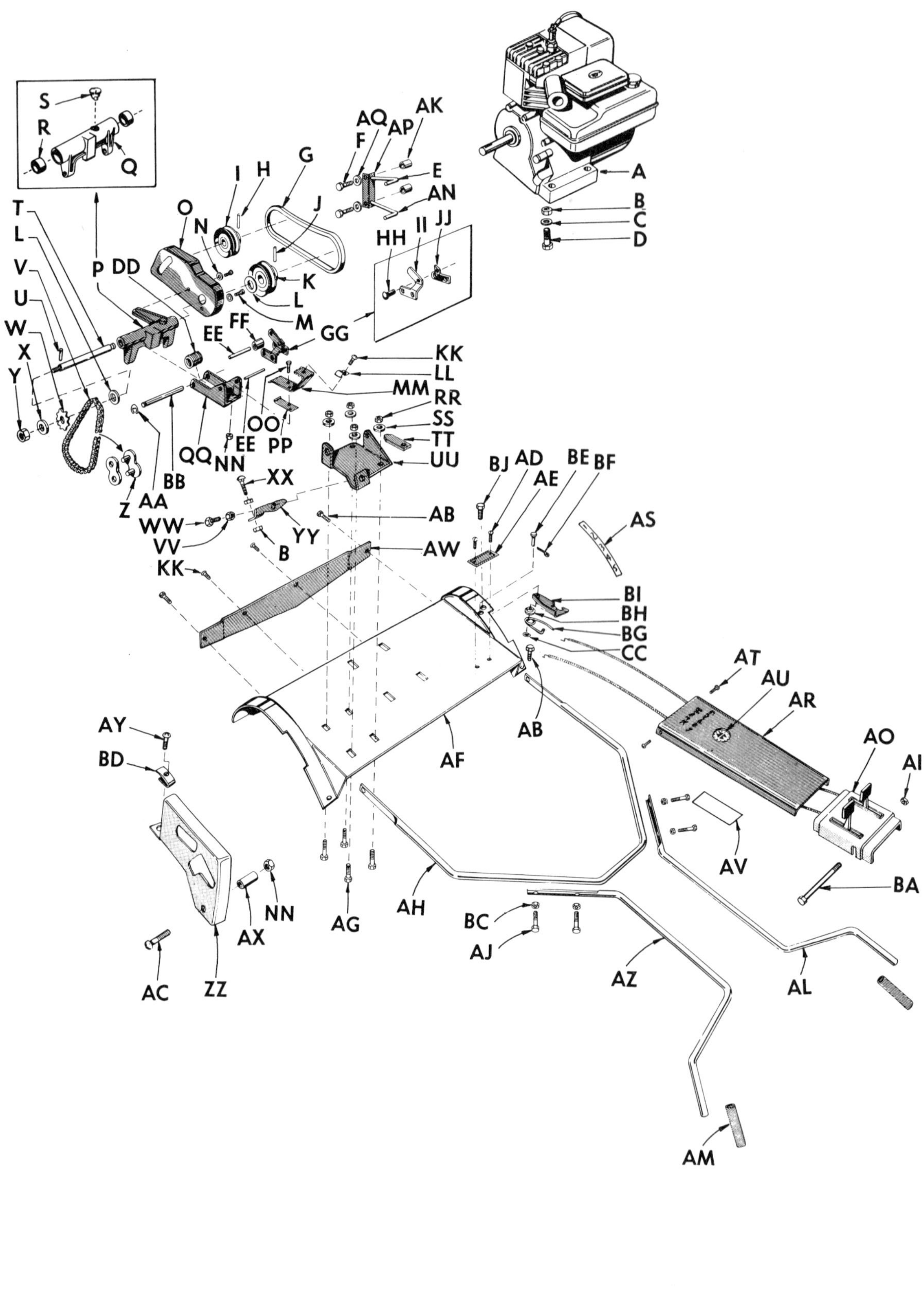

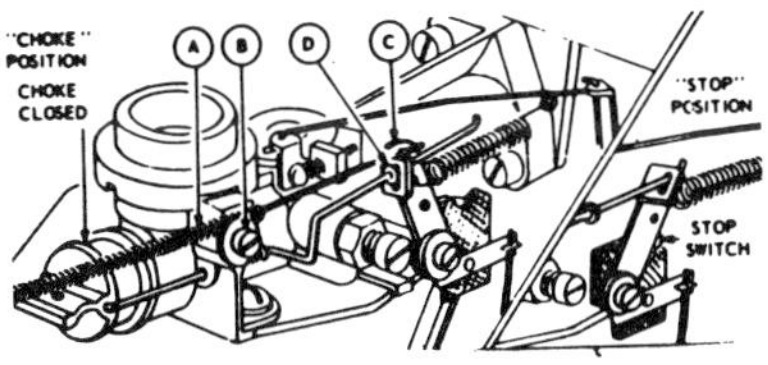

Figure 110

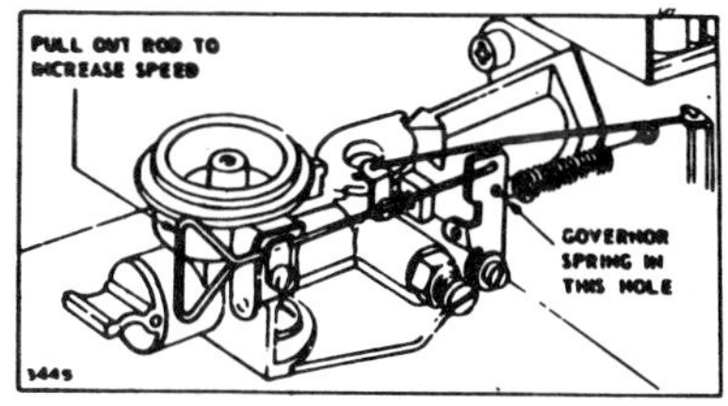

Figure 111

Carburetor Adjustment.

1. Close needle valve clockwise, then back out about 1 1/2 turns.
2. Start engine and set throttle at normal running speed(3000RPM's).
3. Turn needle valve clockwise until engine starts losing speed.
4. Open needle valve slowly, past point of smooth running, until engine just starts running unevenly. Mixture at this point should be perfect for operation under load conditions.
5. Hold throttle in idle position and turn idle adjusting screw to a fast idle of about 1750 RPM's.
6. Test engine under full load. If it stalls, or dies, mixture is still a little lean. Turn needle valve out slightly for richer mixture. Engine may run uneven during idle circuit with the richer mixture.

Air Vane Governor Adjustment.

1. To increase speed, turn adjusting screw clockwise(figure 111).
2. To decrease speed, turn adjusting screw counterclockwise.
3. On manual type governor, adjust per figure 112.
4. To adjust on powered equipment with remote controls, move speed lever in the direction which will elongate governor spring(fig. 113).
5. To adjust, loosen clamp screw on carburetor or fuel tank bracket, and move casing as required to attain proper speed. Maximum speed recommended by the engine manufacturer is 3600 RPM's.

Figure 112

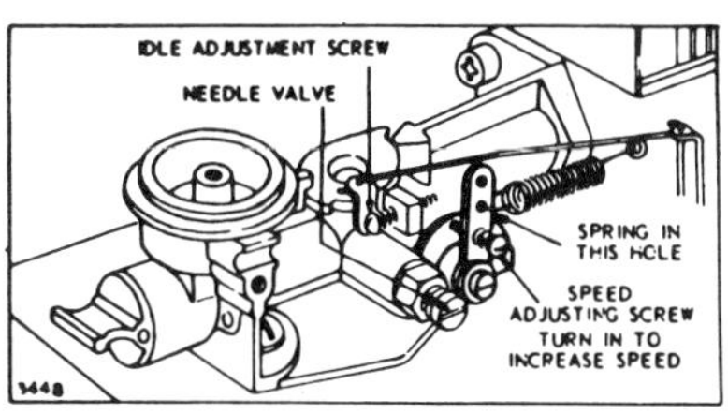

Figure 113

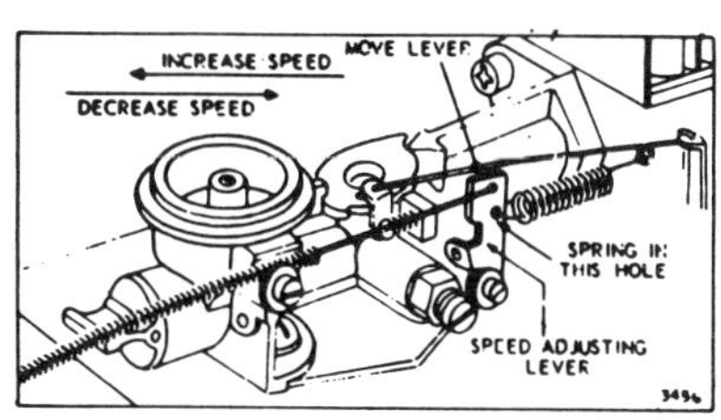

Spark Plug.

1. Use AC type GC46, AL-A7R, or Champion J-8 plugs.
2. The recommended gap on the spark plug is .025".

Figure 114

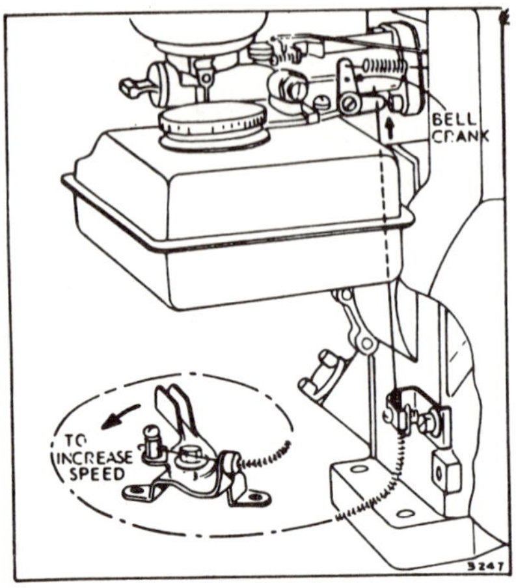

Figure 115

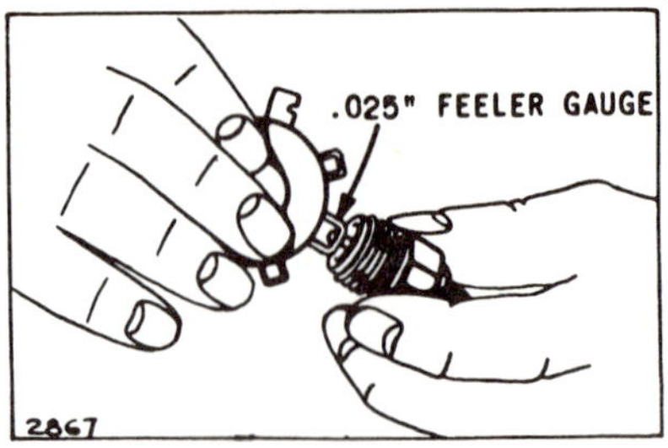

Removing Carbon Deposits.

1. Clean combustion chamber, top of piston, and around both valves.
   a. Service every 100 to 300 hours of operation as required.
2. Remove cylinder head by removing screws(figure 116).
   a. Clean all carbon deposits and foreign matter from parts.
3. Replace cylinder head, washers and screws, in order of removal.
4. Follow sequence in figure 116. Place longer screws where noted.
5. Run engine for 5 minutes then retighten all screws snugly.

Breaker Points.

1. Set breaker point gap at .020".
2. To service points, blower housing and flywheel must be removed.
   a. Rope starter crankshafts have a left hand thread.
   b. Rewind and Wind-up starters have right hand threads.
3. Correct valve clearance.
   a. Intake valve .005" - .007".
   b. Exhaust valve .009" - .011".

Figure 116

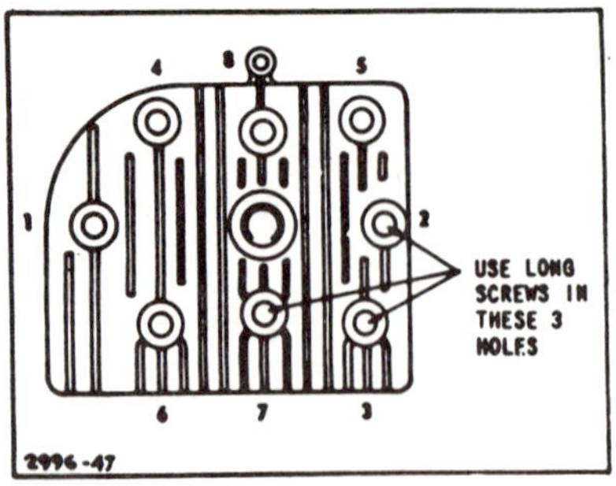

Figure 117

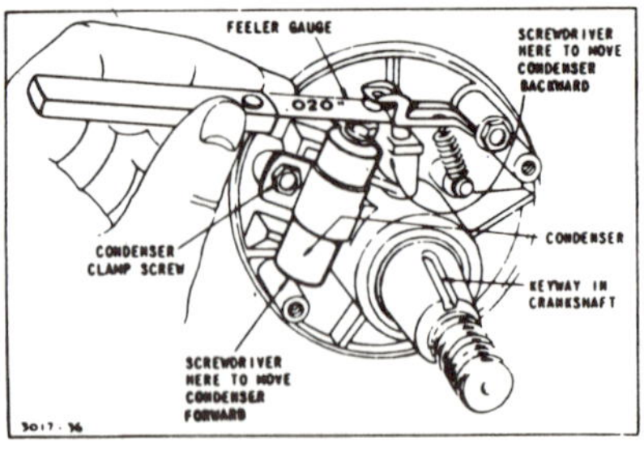

Chapter Six

Jacobsen Manufacturing Company

Jacobsen Manufacturing Company of Racine, Wisconsin is one of the top lawn mower manufacturers in the United States. They make a full line of mowers, and manufacture their own engines as well.

In this chapter we are departing from the normal format and will include the Jacobsen engine in context with the mowers.

Figure 118

Jacobsen has manufactured a number of engines down through the years but for the past eight years, since 1962, have concentrated on the Model 321, in both vertical and horizontal models. Different style engines are utilized on smaller mowers, but both are the 321 Super Torque engine.

The 321 engine is a two cycle unit, as are other Jacobsen engines, and requires a fuel mixture of gasoline and oil. A quick reference back to Chapter Two will give you the operational advantages of the two cycle unit.

Figure 119

Cross section of typical Jacobsen Engine

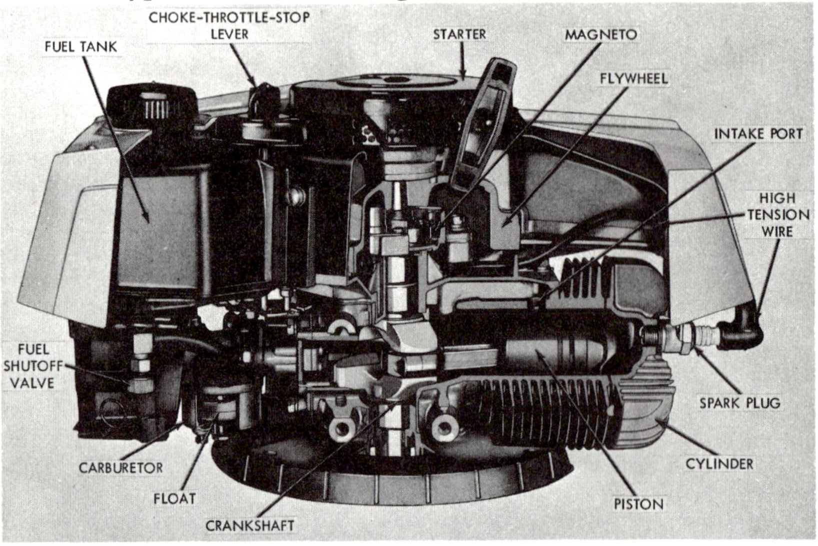

Engine Service and Maintenance.

Lubrication.

1. This engine features an innovation called "Centric-Flo" induction system which controls the fuel and oil flow into the crankcase, and provides proper amounts of fresh lubrication to all moving engine parts with every stroke of the engine. This innovation allows engine to operate on a 32 to 1 fuel mixture, or about 50% less oil than on comparable engines.
2. Fuel must be properly mixed, and fresh, clean material.
3. For best results, keep a mixing can solely for the mower.
4. Use only SAE 30 non-detergent MM oil, and regular gasoline.
5. Pour ONLY 4 ounces of oil into 1 gallon of regular gasoline.
6. Tighten fuel can cap and shake for several seconds.
7. When mixture obtained, pour fuel into mower engine tank.
8. Never add fuel with engine running. Clean away all spillage.
9. On Jacobsen engines prior to the 321, use the standard mixture of a pint of oil to 1 gallon of regular gasoline.
10. Air cleaner service is similiar to models covered previously. Four different types are used, aluminum foil, paper filter, oil bath, and foam. All are subject to clogging and if not serviced will restrict air flow to the carburetor.
    a. To clean foil type, wash in gasoline or solvent, shake off excess cleanser and immerse in SAE 30 oil. Drain off excess and install.
    b. To clean paper filter type, wipe off outer area and tap gently to dislodge dirt from inside. DO NOT immerse in solvent or oil. If filter becomes damaged, or can't be cleaned, replace it.

c. To clean oil bath type, remove cover from oil reservoir and remove all old oil. Wash in gasoline or solvent, and dry thoroughly before refilling to level indicated by arrow. Use same oil used in engine lubrication.

d. To clean foam type, remove filter from case. This style filter should be cleaned in a kerosene-type solution. Allow a few minutes for drying then add 2 to 3 tablespoons of SAE30 oil. Squeeze filter until the oil is evenly distributed throughout the foam.

Starting and Running the Engine.

1. Before starting, drain a small amount of fuel from carburetor bowl. Push up on drain valve. This important step will drain any moisture or foreign matter from the carburetor system.
2. Set fuel control on choke. See fig. 120,or 120a(depending on unit).
3. Fill tank with fuel mixture per previous instructions.
4. Open fuel shut-off valve.
5. If equipped with re-wind starter, grasp starter handle firmly and pull up to start the engine.
6. Keep hands and feet away from cutter blades. See figure 121.
7. When engine starts, move throttle control to FAST, or HEAVY, position. On the Trim-O unit, move choke control back to RUN position.
8. To stop engine, move control to STOP position.
9. Check engine area around cooling fins and keep it clean.
   a. Check air intake screen and keep it free from accumulations of grass, dirt and other foreign matter. Packed dirt and grass around starter and head area will cause overheating which may result in serious damage to the engine.
10. Check carburetor settings occasionally. Overrich mixture may also be injurious to engine and prevent peak performance by engine. A too lean mixture will reduce engine efficiency.

Figure 120

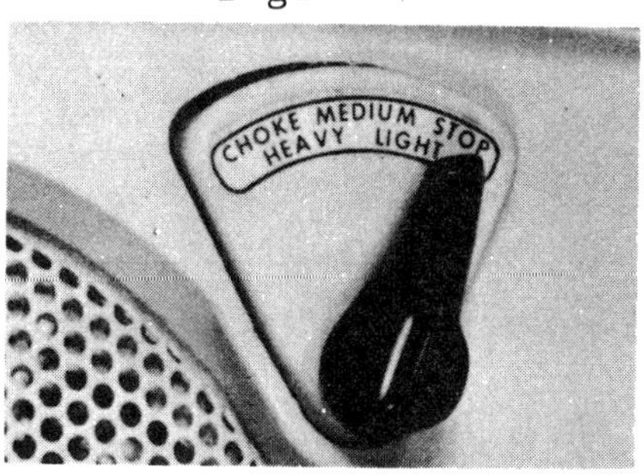

Figure 120a

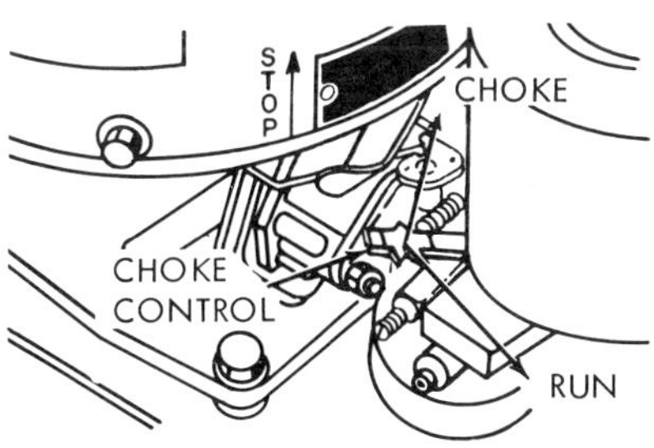

Figure 121

The Carburetor.

Most adjustments to the carburetor should be of a minor nature. Factory settings normally remain in adjustment for long periods of time.

1. Before making an adjustment, clean and service the air filter per previous instructions.
2. To adjust power mixture (controls mixture under load) refer to fig. 122, below.
3. Close power mixture adjusting screw by turning clockwise. Seat the screw gently, don't force. When screw is closed, flat side of screw will align with notch in carburetor body(figure 123).
4. Open one complete turn, using notch for gauge, until flat side of the screw is aligned.
5. Start engine and run wide open for about two minutes.
6. Adjust power mixture adjusting screw to obtain fastest engine speed.
7. Make adjustment in small increments, clockwise. Use simple chart illustrated in figure 124 and make adjustments in 1 or 2 hour steps.
8. If you overadjust, and engine speed starts decreasing, turn back the opposite direction as needed. Always allow engine at least 2 minutes of running time at each setting to establish a pattern.
9. If engine stalls under load, enrich mixture slightly by turning adjusting screw back, or counterclockwise one unit(1 hour). The flat side of adjusting screw should never be more than 3 hours away from notch on either side. When setting falls within this range it should function properly. If not, additional service is required.
10. Do not try to regulate speed by adjusting carburetor mixture.
11. When control is set on CHOKE position, choke valve should resemble illustration in figure 126.
12. After power adjusting screw is set, idle mixture screw can be adjusted properly.
13. Keep engine running and move throttle control back to SLOW.
14. Turn idle adjusting screw inward(clockwise) in 1 hour increments, until fastest idle speed is reached.
15. Turn adjusting screw back about 2 increments(hours), or until engine is running smoothly.
16. To set idle speed after adjustment(engine will be idling at fast rate) turn idle adjusting screw back(counterclockwise) until idle speed is set.

Figure 122

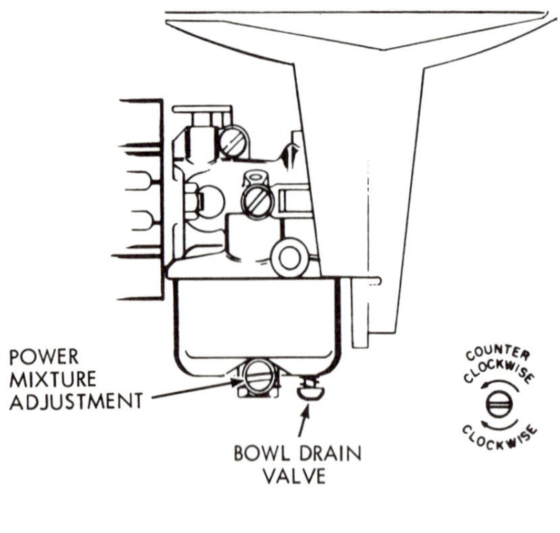

Figure 123

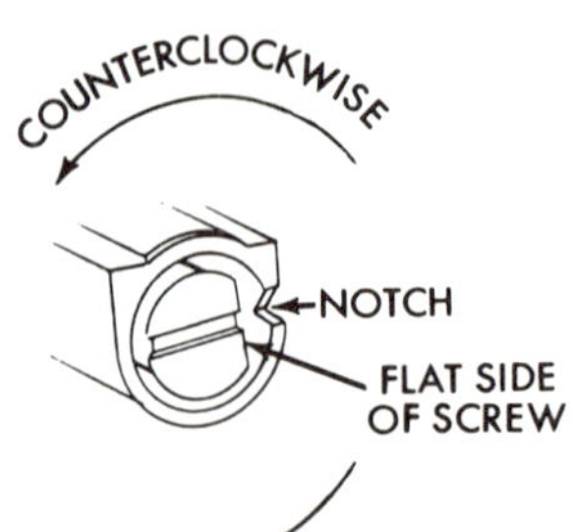

Figure 124
Clock increments for adjustment

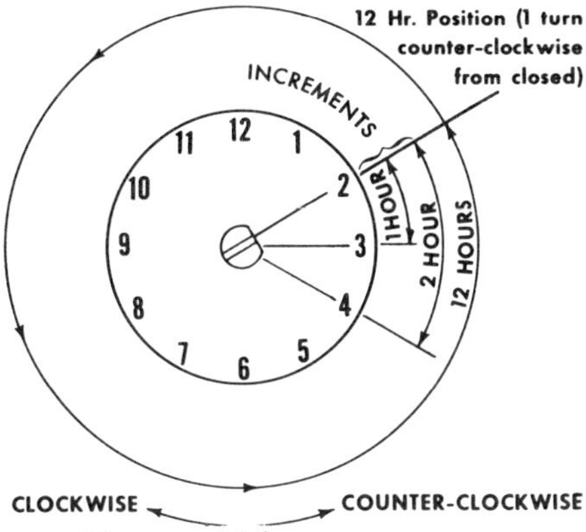

Figure 125
Maximum adjustment setting-Walbro

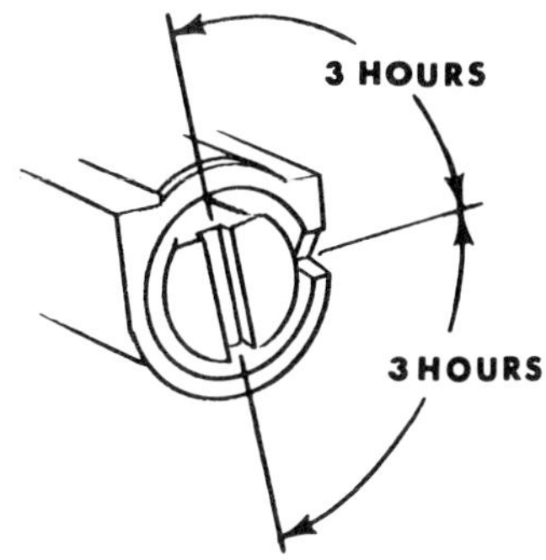

Figure 126

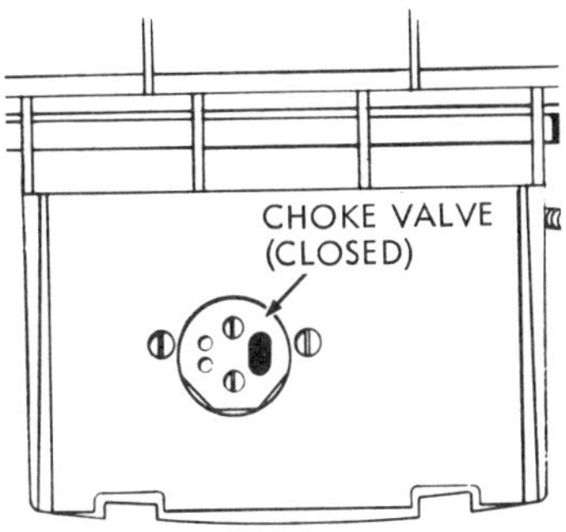

Figure 127

Carburetor Overhaul.

1. If a major adjustment or overhaul is required, carburetor must be removed from engine, after removing fuel line from inlet(fig. 127).
   a. If removal is difficult, use a screwdriver and pry off.
2. Remove wing nuts which hold air cleaner to carburetor.
3. Remove four screws that hold carburetor to engine.
4. Disengage governor linkage, remove carburetor/adapter assembly.
5. Before disassembling the carburetor, turn to page 84 and familiarize yourself with the components, which are numbered in sequence, in order of removal.

Figure 128

Figure 129

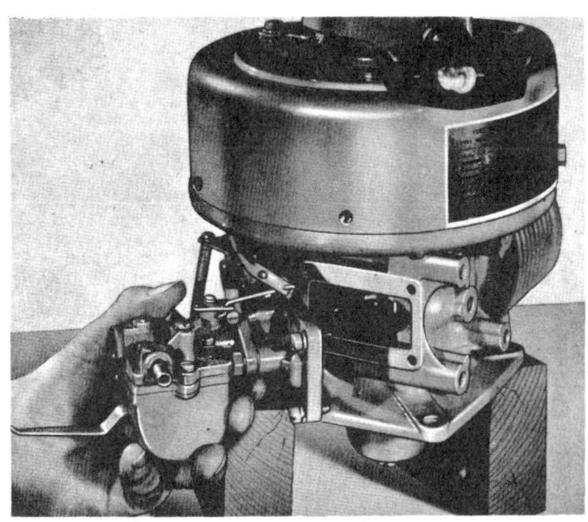

Figure 130
Carburetor components numbered in order of removal for service/overhaul

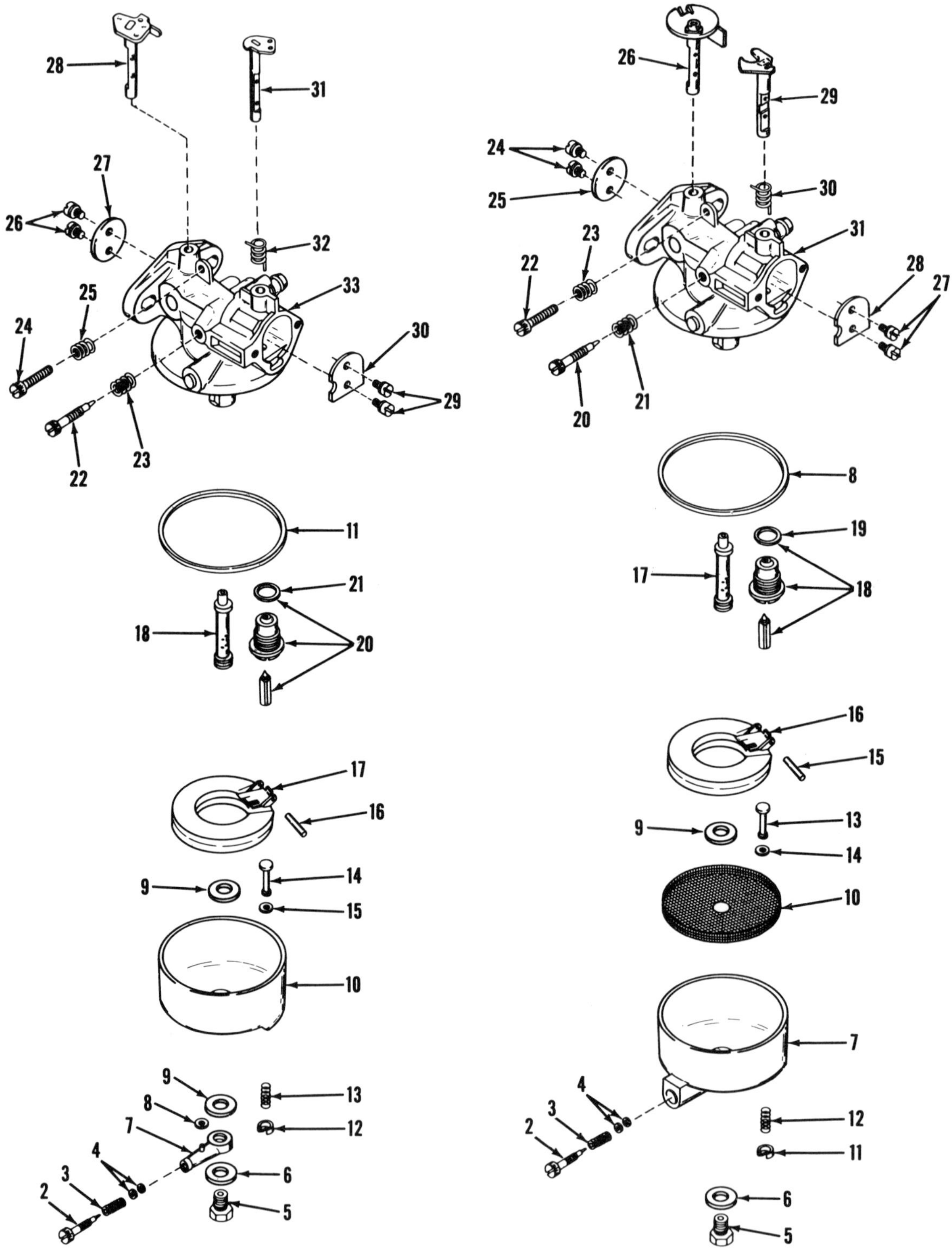

6. After disassembling the carburetor, clean all components in a good carburetor cleaner. Cleaner is available from your local automotive jobber, and contains properties designed to remove caked residue that forms in the carburetor. Follow manufacturers recommendations.
7. Inspect needle ends of idle and power adjusting screws. If bent or scored, replace them. If they are damaged, also inspect needle valve seats, which will probably be damaged. Dress smoothly if possible.
8. Inspect all gaskets. If any are slightly marred, bent, or broken, replace them. If engine has had prolonged usage, replace all gaskets.
9. Inspect reed valve surface of adapter. The surface must be absolutely clean and free of damage. Reed must be installed to completely cover intake hole and be perfectly flat against the adapter. Reed has a slight bend to apply tension against the adapter and function as an inlet valve. If surface isn't clean, or reed is improperly positioned, blowback through the carburetor will result.
10. Reassemble components in reverse order of removal.
11. Use new packing for main adjusting screw. Thread new packing in place with black face away (toward you) from needle. Thread on until three or four threads are exposed on adjusting screw, then install screw into carburetor until needle point is barely seated.
12. Install packing nut until it's tight against packing, and back adjusting screw away from seat 1 1/4 to 1 1/2 turns.
13. To set carburetor float on a Tillotson unit, use idle tube as gauge for proper setting of 1/16-3/32" between bowl cover flange and nearest edge of float, when bowl cover assembly is in an inverted position. To adjust, bend tab on float. If unit utilizes a plastic float, use same procedure, but set the distance at 5/16". See figure 131.
14. To set carburetor float on Walbro unit, hold carburetor in an inverted position.
15. Adjust clearance between float and casting rim to 5/32" by bending tab(A), figure 132.
16. Float travel setting(see fig. 132) is 3/16". Bend tab(C) as needed.
17. Carburetor adjustments may be made by turning back to page 83.
    a. The basic concept above can be applied to most carburetors.

Figure 131

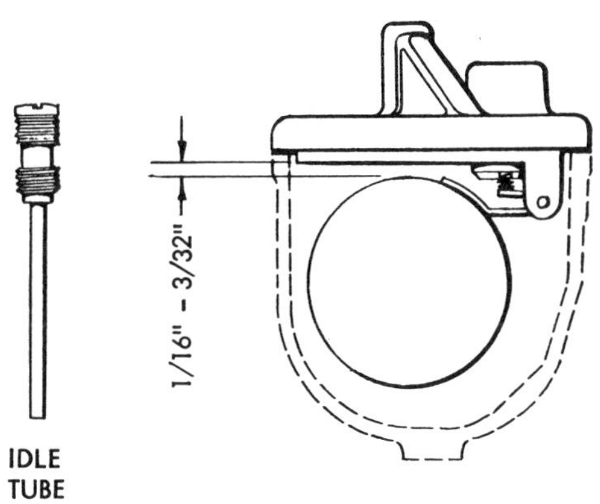

Figure 132

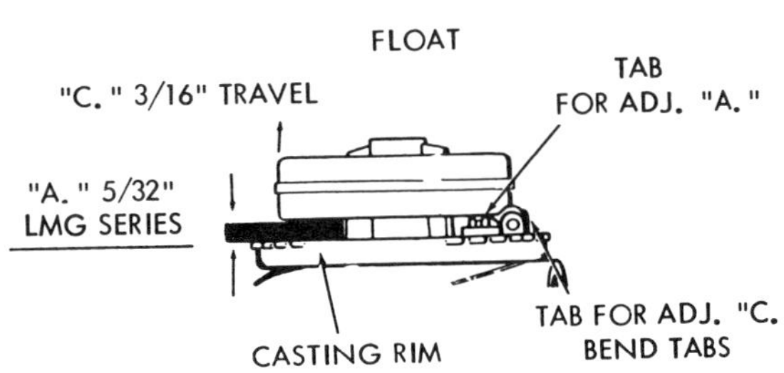

Governor Adjustment.

1. Model 321 engines with positive control have an adjusting screw on the lever which links the positive control lever to the governor.
   a. To increase engine speed, under governor control, turn screw clockwise to increase governor spring tension.
   b. To decrease engine speed, turn screw counterclockwise.
   c. If possible, use a vibration tachometer to check speed in accordance with manufacturers specifications on page 88.

Choke Adjustment.

1. Choke adjustment is necessary if handle or control wires have been removed or altered.
2. If mower has the single "choke-fast-idle-stop" control on handle, set control in FAST position. The hook in choke link should either contact or be 1/16" away from the first loop on choke spring. See figure 134.
3. To adjust, loosen throttle housing screw and move housing in proper direction to correct, and tighten the screw securely.
4. Be certain engine control cable is properly installed on carburetor. Controls on handle must correspond with settings at carburetor. When control is moved to CHOKE position, choke on carburetor must be fully closed. When control is moved to STOP position, engine control lever must make solid contact with spring STOP switch. See figure 135.
5. If adjustment is required, loosen clamp screw holding control cable and slide cable back and forth. To set, move control lever on handle to FAST position, and move cable until choke control lever barely touches carburetor choke, with choke completely open. Tighten clamp screw.
6. To test the adjustment, move engine control lever to STOP position. If choke control lever contacts spring STOP switch, and carburetor is fully choked when lever is moved to START position, adjustment is correct.

Figure 133

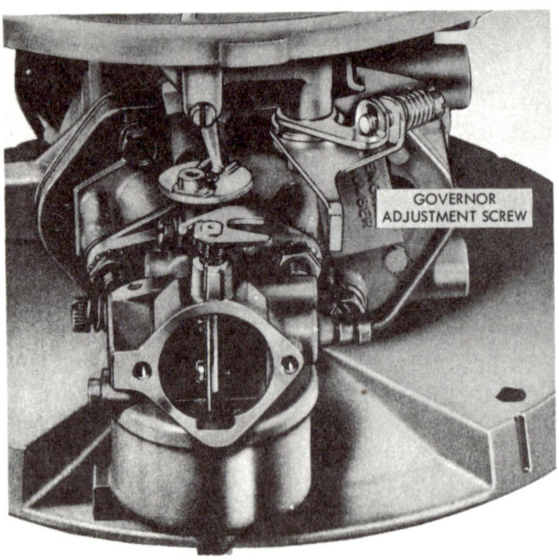

Figure 134

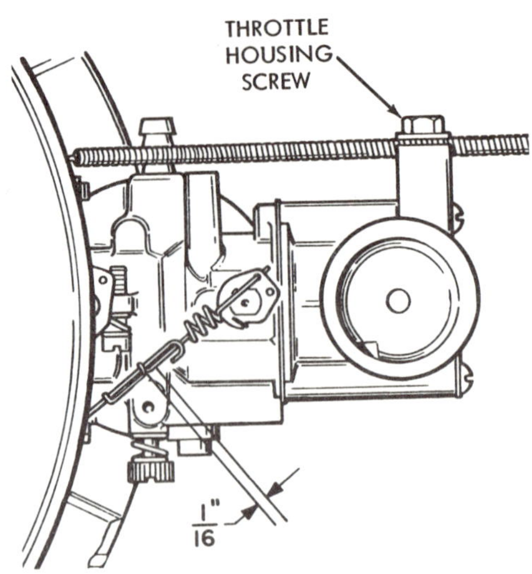

Figure 135

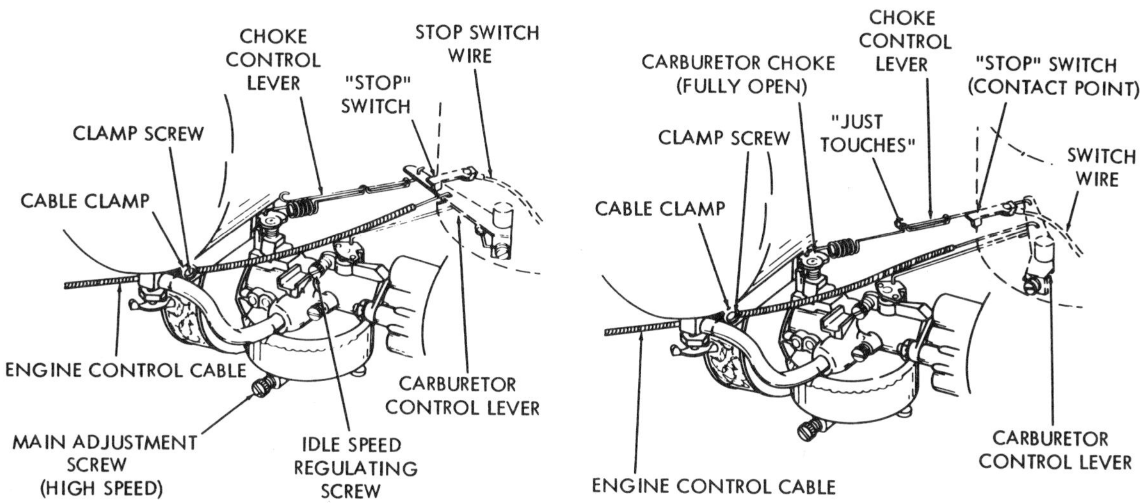

"STOP" POSITION

FAST (HIGH SPEED) POSITION

The Spark Plug.

1. The spark plug should be inspected, regapped to .030", and cleaned thoroughly, every 100 hours of operation. Since spark jumps the gap nearly 3000 times a minute while engine is running, the plug must be in good condition.
2. Improper fuel, fuel mixture, oil additives are major causes of fouled plugs.
3. Plugs for model 321 engine can be cleaned with Champion sand blast cleaner. Always clean thoroughly with a solvent and use air gun to blow away all sand particles. If the above system is not available, use steps for cleaning spark plugs as outlined in Chapter Two.
4. When replacing the plug, always tighten securely.
5. Refer to spark plug chart on page 34 for proper plug.
6. To check for spark, be certain hooked end of contact spring is pushed through top of spark plug wire and also through high tension wires, to insure positive contact.
   a. Before inserting spark plug wire with spring into cover, apply a liberal soap and water solution inside the cover. Insert wire into cover and be sure spring is all the way into cover and that opening in the spring aligns with opening in cover.
   b. Push cover in place over the spark plug.
   c. To remove cover from plug, pull and twist off at same time.

Figure 136

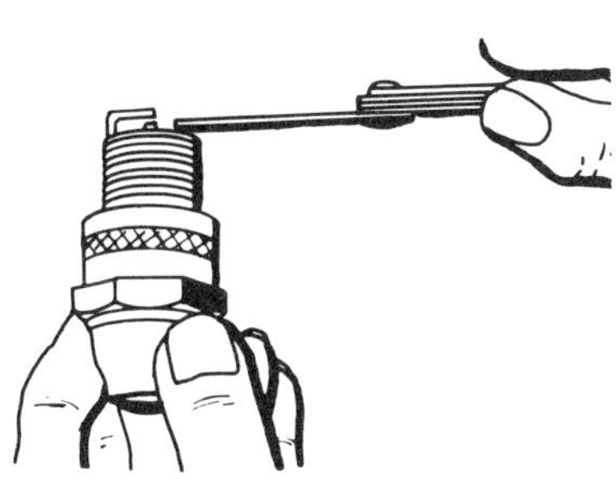

Figure 136a

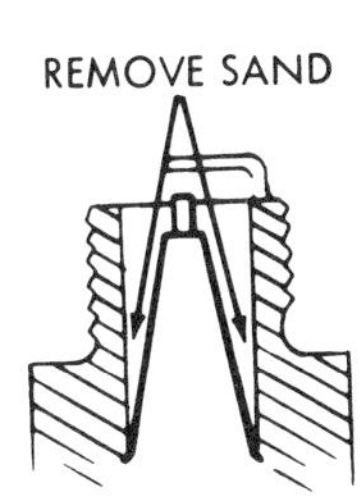

Figure 137
The Speed adjustment chart for Jacobsen line of lawn care equipment.

	RPM Idle	RPM Top Speed
9" Edge-R-Trim (32A9, 32B9, 32C9)	1500 to 2000	3000 to 3300
9" Edge-R-Trim (50012)	1500 to 1800	Up to 3600
10" Trimo (3110 - 8610, 86A, 86B	1500 to 2000	3400 to 3600
10" Trimo (50035)		3500 to 3700
18" Pacer (52C18, 42D18, 42E18)	1300 to 1600	Up to 3500
18" Pacer (11814)	1500 to 1800	Up to 3600
18" Turbo-Cut (3418, 34B18, 34C18, 34D18)	1300 to 1800	5000 to 3200
18" Turbo-Cut (7518, 75A18, 75B18)	1500 to 2000	3400 to 3500
18" Turbo-Vac (31817)	1500 to 1800	3200 to 3400
18" Turbo Cone (117-18)	1500 to 1600	3200 to 3400
20" Scepter (8020, 80A20)	1500 to 1800	3000 to 3200
20" Commercial Rotary	1500 to 1600	3200 to 3400
20" Snow Jet (9620, 96A20	1500 to 1800	3600 to 3800
20" Snow Jet (52002)	1700 to 1900	3600 to 3800
21" Turbo-Cut (3921, 39B21, 39C21)	1500 to 1800	3200 to 3400
21" Turbo Cone (119-21)	1500 to 1600	3200 to 3400
21" Turbo-Cut (3521, 35C21, 35D21, 35E21, 35F21)	1500 to 1800	Up to 3500
21" Turbo Cone (121-21)	1500 to 1600	3200 to 3400
21" Lawn Queen (2C21, 2D21, 2E21)	1300 to 1600	Up to 3500
21" Lawn Queen (12113)	1500 to 1800	Up to 3600
21" Manor (28F21, 28G21)	1300 to 1600	Up to 3500
21" Manor (22114, 32121-7B1)	1500 to 1800	Up to 3600
22" Putting Green (9A22, 9B22)	1500 to 1800	Up to 3400
22" Greensmower (62203)	1500 to 1800	Up to 3800
22" Scepter (80A22)	1500 to 1800	3000 to 3200
24" Estate (8A24, 8B24)	1300 to 1600	Up to 3800
24" Rotary S.P. (40A24)	1900 to 2100	Up to 3750
26" Estate (8A26, 8B26, 8C26, 8D26)	1300 to 1600	Up to 3800
26" Estate R.R. (22601, 22605-7B1)	1500 to 1800	Up to 3800
26" Estate F.R. (22611, 22615-7B1)	1500 to 1800	Up to 3800
26" Lawn King (12A26, 12B26)	1300 to 1600	Up to 3200
26" Lawn King (12601)	1500 to 1800	Up to 3800

	Light Setting	Medium Setting	Heavy Setting
18" 4-Blade Rotary (31809)	2400 to 2600	2800 to 3000	3200 to 3400
20" Commercial (32028)	2400 to 2600	2800 to 3000	3200 to 3400
20" Robust (32031)	1500 to 1700		3200 to 3400
21" 4-Blade Rotary (32114)	2400 to 2600	2800 to 3000	3200 to 3400
21" 4-Blade Rotary S.P. (42114)	2400 to 2600	2800 to 3000	3200 to 3500

Ignition-Magneto.

1. For easy starting, and good performance, set spark plug gap at .030" and magneto breaker points at .020".
2. When magneto points are set too close (.008" to .010") mower may be hard to start. When points are set too wide (.030" to .032"), plugs will foul easily, and secondary voltage output will be erratic.

The magneto is a stationary self-contained generating unit consisting of an armature plate assembly, condenser and breaker point assembly.

The permanent magnet built into the flywheel revolves around the stationary pole pieces in the armature plate assembly. As the ends of the magnet pass by the pole pieces, an alternating magnetic flux is established through the ignition coil. A current is generated in the primary circuit during the period breaker points are closed. A cam located on the crankshaft opens the points when the primary current is at a maximum. Interruption of the current breaks the magnetic field around the coil. The collapse of the magnetic field induces a high voltage through the secondary winding of the coil, which is then carried through the high tension wire to the spark plug. The current then jumps the spark plug gap to ignite the fuel charge in the cylinder. Each revolution of the crankshaft repeats the cycle of ignition. The condenser, connected across the contacts, absorbs the current which tends to continue after breaking contact. This speeds the collapse of the magnetic field and reduces pitting of breaker points by absorbing the sparking across the points.

Figure 138
Open view of magneto and components

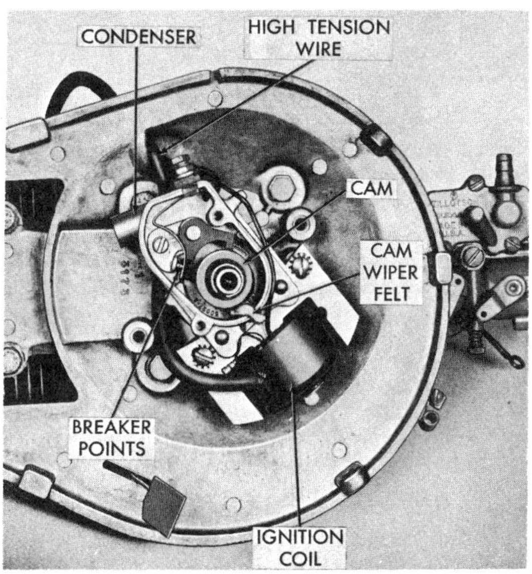

Figure 139

1. Before dismantling engine to service magneto, check to make certain the problem is there. If a test plug unit is not available, remove spark plug wire and hold next to head. Crank engine and watch for spark. If spark is getting through wire, problem is in spark plug. If no spark is getting through, proceed as follows.
2. Remove the flywheel and inspect breaker points(figure 138).
3. If pitted and corroded, replace with new points.
4. After setting points, clean again by drawing a piece of card stock between them.
5. Check stop switch wire for breaks, or short-outs at terminal on the tank panel or carburetor throttle control. Check all wires for looseness or short-outs.
6. Replace the flywheel and test again.
7. If spark is still weak or non-existent, check the coil and condenser. Special test equipment is required for this inspection. If the parts are faulty, replace them.

Removing the Flywheel.

1. Remove the starter and fan housing from engine. Figure 139.
2. Remove screws from flywheel housing, high tension wire clip and slip off housing. If model features attached fuel tank, remove it, after disconnecting the fuel line and carburetor linkage.
3. Remove spark plug and if available, use special flywheel holder tool and insert in opening.
4. Turn flywheel until piston is bottom dead center. Screw stud of tool inward until it locks piston, and tighten the lock nut.
   a. Use a 5/8" socket on nut of starter pawl assembly and remove. See figure 140.
5. Screw "knock-off" nut down fully on the crankshaft end.
6. Rap top of nut sharply with a hammer to jar flywheel loose on taper of crankshaft. Figure 141.
7. Remove the nut and flywheel from crankshaft.
8. If a knock-off nut isn't available, screw original nut back on crankshaft until top of nut is flush with end of shaft. Place a hardwood block on top of the nut and rap block sharply to jar flywheel loose.
   a. DO NOT pry on the underside of flywheel to loosen.
   b. DO NOT lose key from crankshaft keyway.
9. Never strike crankshaft directly with hammer. Always use either a knock-out nut, or a block of wood to prevent damage to threads.

Figure 140

Figure 141

Replacing Breaker Points and Timing the Magneto.

1. Remove cover plate and gasket from stator assembly. This would be accomplished if you had inspected the points.
2. To replace the points, remove connection stud nut, connection stud and contact clamp screw.
3. If points are badly pitted, save another service situation and replace the condenser at the same time.
4. Flywheel requires no recharging since magnets hold indefinite charge.
5. Check all soldered connections. Resolder if necessary.
6. Check breaker spring and mechanism.
7. Check coil and condenser per previous instruction.
8. Replace cam wiper felt and cover with soft grease.
9. To set breaker points, turn crankshaft until breaker arm shoe is on high point of cam. If you have a feeler gauge, set at .020".
10. To adjust points for proper setting, loosen adjustment screw with a screwdriver and move breaker plate as needed. Figure 142.
11. Tighten screw and recheck gap setting.
12. The ignition timing is fixed and requires little or no adjustment. If the stator plate has been removed, or if two fillister head screws securing the stator plate through the timing slots have been loosened, rotate the plate fully, clockwise, before tightening the screws.
13. Recheck breaker gap setting.

If complete removal of magneto is required, proceed as follows.

1. Remove flywheel key from the crankshaft.
2. Remove the two fillister head screws holding stator plate assembly to magneto back plate.
3. Remove magneto and pull high tension wire through the back plate.
4. Remove cam positioning ring from crankshaft(figure 144).
   a. Use a screwdriver to pry up. Don't damage the keyway.
5. Remove four screws holding back plate to crankcase(figure 145).
6. Refer to figure 146 for dis-assembly steps. Remove in numerical-order to magneto unit (34), or remove key (35) step one and proceed to step five and remove back plate from crankcase.

Figure 142

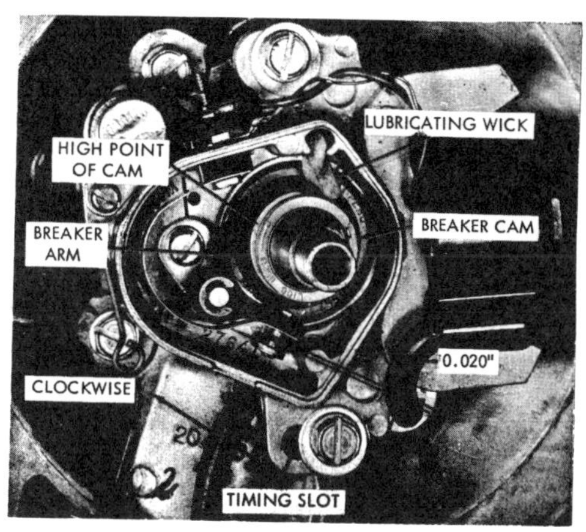

Figure 143

Figure 144

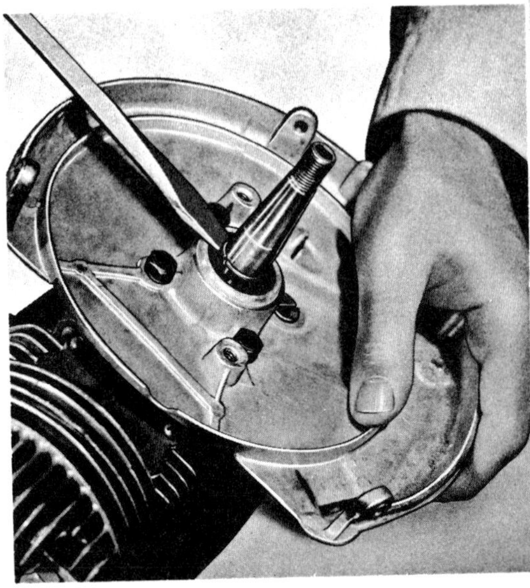

Figure 145

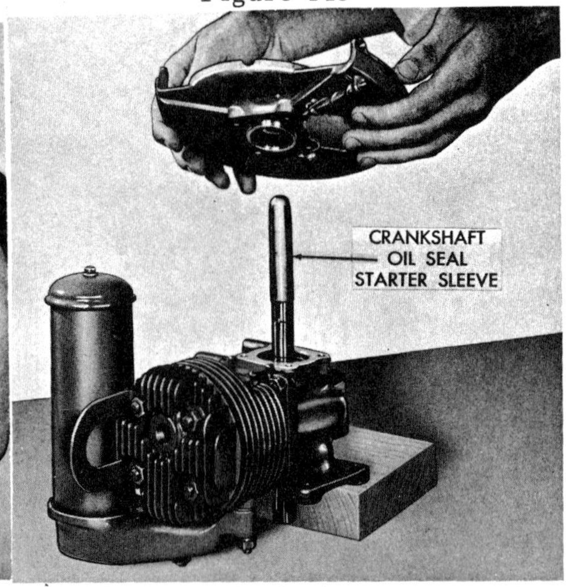

7. Reassemble with special emphasis on following items. Reverse order of removal relative to numbered sequence.
   a. High tension wire should be pushed through magneto back plate.
   b. Smaller diameter of breaker cam faces upward.
   c. Magneto points and timing are properly adjusted.
   d. Key must be inserted into crankshaft before installing flywheel.
   e. Nut on starter pawl assembly must be tightened securely.
8. Check engine operation after assembly.

Figure 145a

The illustration below is parts breakdown and description of Wico magneto assembly. Numbers are used to identify parts only.

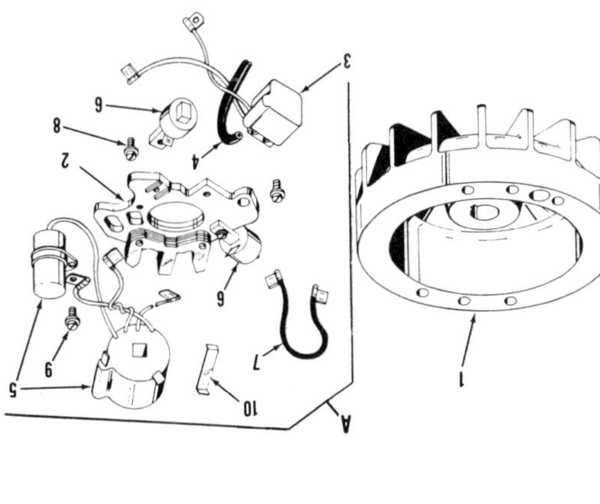

Ref. No.	Part No.	Description
	156879	Wico Magneto Assembly, Model No. FW2901, consists of the following Ref. Nos.
A	500697	Stator Plate Assembly, (consists of Ref. No's 2 - 10)
1	204227	Rotor
2	500680	Stator Plate
3	545809	Coil, Output
4	325791	Lead Wire, High Tension
5	500681	Charging Coil and Condenser Assem
6	545810	Trigger (2 required)
7	545812	Wire, Trigger Interconnecting
8	407574	Screw, Coil and Trigger Mtg
9	407702	Screw, Condenser Mounting
10	314462	Spring, Coil Retaining

Figure 146

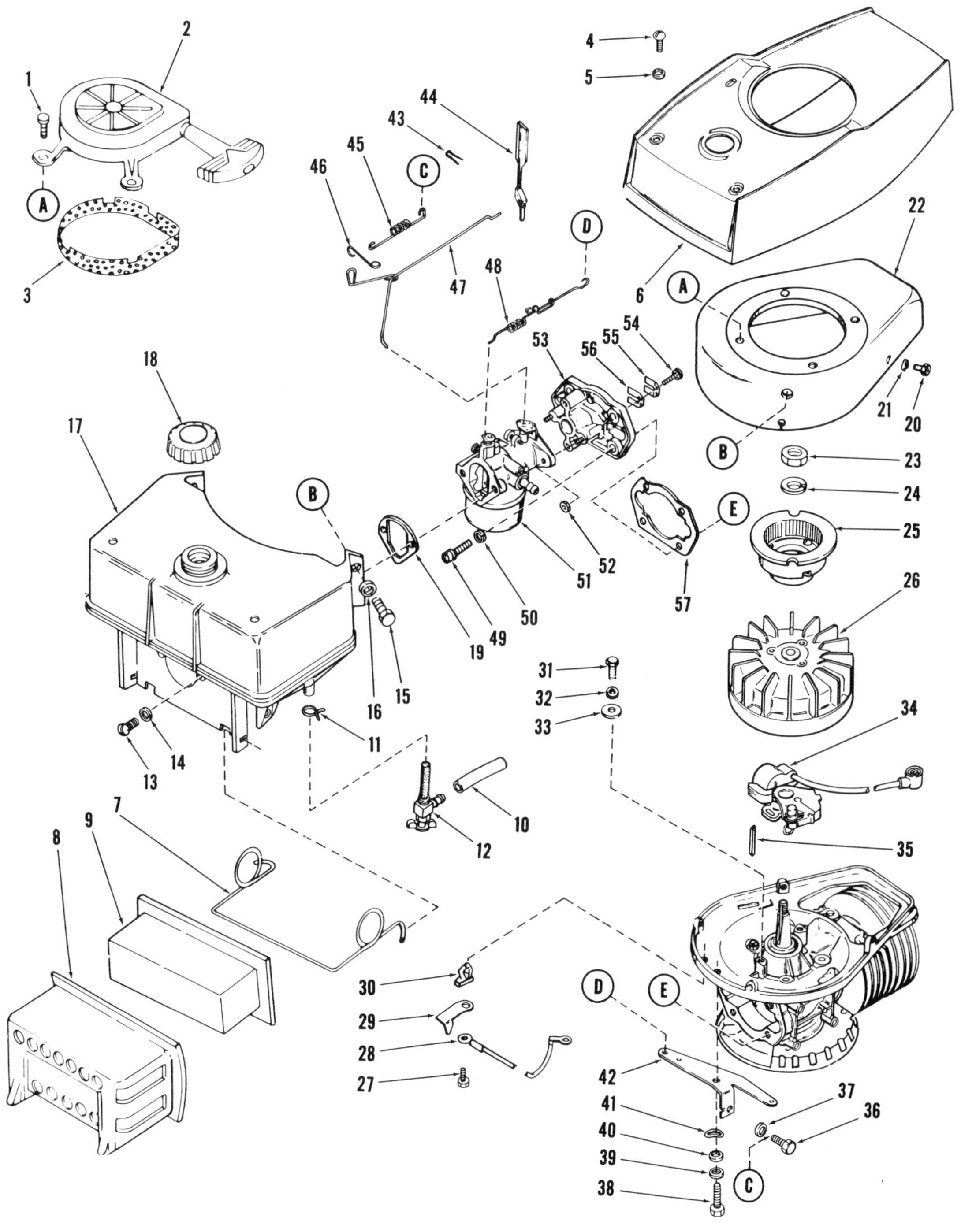
1
2
3
A
4
5
6
22
A
B
C
D
E
7
8
9
10
11
12
13
14
15
16
17
18
19
20
21
23
24
25
26
27
28
29
30
31
32
33
34
35
36
37
38
39
40
41
42
43
44
45
46
47
48
49
50
51
52
53
54
55
56
57

Fuel Line and Tank.

1. To clean the fuel line, remove from inlet on the carburetor.
2. Refer to figure 146, the number sequence starting at the fuel tank, and remove other end of fuel line, and fuel tank if necessary.
3. Blow out air line to remove obstructions and clean with grease solvent.
4. If necessary, clean the tank using the same solvent.
    a. Remove shut-off valve to clean it and the screen.

Starter Service.

1. To overhaul automatic rewind starter unit, begin by pushing cord through the grip, untie the knot and remove the grip.
2. Release cord into starter slowly so the spring won't unwind suddenly.
3. Remove roll pin and washers from ratchet shaft. Remove the pulley
4. Remove spring from pulley and bracket and remove cord from pulley.
5. If cord is broken or frayed, replace. If cord replacement is only service required, reassemble the unit from that point.
6. Replace pawl springs if necessary.
7. Install recoil spring by hooking ends of spring in extruded hook in bracket and replace pulley in bracket. Install washers and roll pin.
8. Wind pulley counterclockwise about 4 turns(small starter), or 4-3/4 turns on the large starter.
9. Clamp starter in vise with cord in position illustrated in fig. 147.
10. Insert cord through eye inside of pin, with open side of clip on cord facing outward, and thread the cord through the sleeve.
11. Insert through grip. Put cord retainer in place, if removed. Tie knot in cord and pull into grip firmly to seat the knot.
12. Replace assembly on the engine and test.

Figure 147

Figure 148

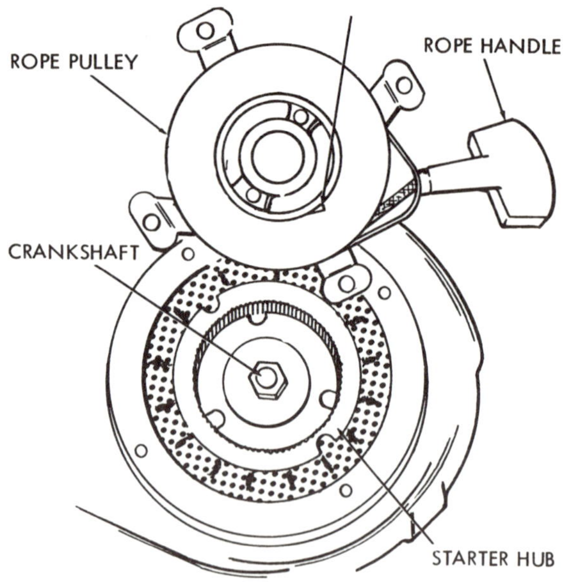

1. To service Eaton starter, remove four screws holding assembly to engine fan housing and lift off.
2. Before disassembling, release tension of rewind spring. Fig. 149.
3. Hold starter assembly with pulley up. Pull on starter rope until the notch in pulley is aligned with hole in rope bushing. Press on starter with thumb to prevent pulley from rotating.
4. Feed rope through rope bushing in reverse direction to obtain slack between pulley and rope bushing.
5. Lift slack portion of rope until it engages notch in pulley.
6. Release thumb pressure slowly and allow spring to unwind until all tension is released.
7. Remove starter pulley by removing flat head screw from center shaft.
   a. When removing pulley, keep spring confined in the housing.
8. If rope is damaged, remove screws holding pulley halves and replace.
9. To reassemble, first install a new spring if needed. Keep spring confined in housing. Replacement springs are provided with holders to aid assembly. The holder slips off as spring is inserted in housing. Be certain spring is installed with windings in same direction as old spring.
10. Replace starter rope by crimping between two pulley halves of the starter pulley and tightening screws that join pulley halves. Insert the free end of rope through rope bushing in housing and through handle.
11. Tie a double knot in end of rope and wind rope onto the pulley.
12. To install pulley in housing, align notch in pulley hub with hook end of spring. Use wire bent to form a hook to move spring into correct alignment with hub on underside of pulley. When properly aligned, push pulley part way down into place. Remove wire hook and seat pulley in housing.
13. To place rope under tension, align notch in pulley with rope bushing in housing. Feed rope through rope bushing and hook one finger through loop in slack portion of rope. Pull rope up to engage notch in pulley. Hold rope in position and rotate pulley at least two full turns in same direction as it is pulled, to properly pre-tension the spring.
14. After pre-tensioning spring, pull rope to fully extended position. Spring should be pre-tensioned so it fully reqinds rope when handle is released.

Figure 149

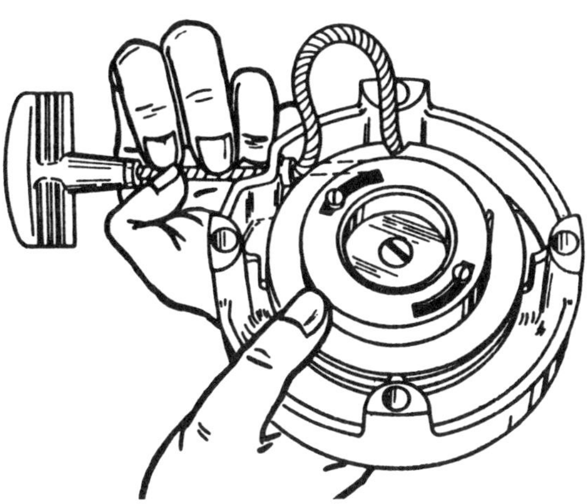

Figure 150

Figure 151
Service Chart Covering Tolerance and Clearance settings for Model 321.

	J-125-H J-100-H	J-125-V	J-175-H J-175-V	J-225-V	J321 Bronze Rod	J-321 Roller Bearing Rod
Cylinder Bore	2.0015 2.0010	2.0015 2.0010	2.1265 2.126	2.2515 2.2510	2.1265 2.1260	2.1265 2.1260
Piston Skirt Diameter	1.9970 1.9965	1.9970 1.9965	2.1240 2.1235	2.2480 2.2475	2.1232 2.1225	2.1232 2.1225
Piston Ring Width	.0905 .089	.0905 .089	a. .0625 .0615	.0925 .0935	.0925 .0935	.0925 .0935
Piston Pin Diameter	.50025 .49975	.50025 .49975	.50025 .49975	.50025 .49975	.5001 .4999	.5001 .4999
Piston Pin Bore in Piston	.5003 .5000	.5003 .5000	.5003 .5000	.5003 .5000	.5001 .4998	.5001 .4998
Piston Pin Bore in Rod	.50125 .50075	.50125 .50075	.5008 .5015	.687 .688 *	.5008 .5015	.5008 .5015
Crank Pin Diameter	.6245 .6240	.6235 .6230	.7485 .7490	.7503 .7500	.7485 .7490	.7501 .7496
Crank Pin Width	.817 .812	.817 .812	.938 .933	.568 .563	.938 .933	.568 .563
Conn. Rod Bore	.627 .626	.627 .626	.7525 .07530	.9403 .9399 *	.7525 .7530	.9819 .9824 *
Conn. Rod Crankpin Width	.808 .800	.808 .800	.929 .921	.555 .550	.929 .921	.558 .550
Spark Plug Gap	.030	.030	.030	.030	.030	.030
Mag. Point Gap	.020	.020	.020	.020	.020	.020
Timing in Degrees B.T.D.C.	30°	30°	28°	27°	26°	None
Timing in Inches of Piston Travel B.T.D.C.	.125	.125	.125	.125	Full Retard	Full Retard
Piston Skirt to Cylinder Clearance	.004 .005	.004 .005	.002 .003	.003 .004	.0028 .0040	.0028 .0040
Maximum Wear Tolerance on Cylinder Bore	.002	.002	.002	.002	.002	.002
Maximum out of round Tolerance	.0015	.0015	.0015	.0015	.001	.001
Piston Ring End Gap	.005 .010	.005 .010	59-.005-.013 60-.052-.060	.005 .013	.005 .013	.005 .013
Piston Ring to Land Side Clearance	.0035 .0055	.0035 .0055	b. .004 .006	.002 .0035	.0015 .0035	.0015 .0035
Piston Pin Clearance in Rod	.0005 .0015	.0005 .0015	.0005 .0017	Needle Brgs.	.0007 .0017	.0007 .0017
Conn. Rod Clearance	.0015 .003	.0025 .004	.0035 .0045	Needle Brgs.	.0035 .0045	Needle Brgs.
Conn. Rod Crankpin Side Clearance	.004 .017	.004 .017	.004 .017	.008 .018	.004 .017	.005 .018

Figure 152
Torque Data Chart--Use whenever attempting major service work.

	J-125	J-175	J-225	Bronze Rod J-321	Roller Bearing Rod J-321
Spark Plug - 180 to 200	Size / Inch Pounds Torque	Size / Inch Pounds Torque	Size / Inch Pounds Torque	Size / Inch Pounds Torque	Size / Inch Pounds Torque
Cylinder Head Bolts or Nuts	5/16-24 / 150-180	5/16-24 / 150-180	5/16-18 / 150-180	5/16-24 / 150-180	5/16-18 / 150-180
Crankcase Head Bolts	5/16-18 / 80-100	5/16-18 / 80-100	5/16-18 / 80-100	5/16-18 / 80-100	5/16-18 / 80-100
Back Plate Screws	1/4-20 / 60-80	1/4-20 / 60-80	1/4-20 / 60-80	1/4-20 / 60-80	1/4-20 / 60-80
Connecting Rod Screws	#12-28 / 65-75	#12-28 / 65-75	#12-28 / 65-75	#12-28 / 65-75	10-32 / 45-50
Carburetor Adapter Screws	#12-24 / 30-50	1/4-20 / 30-50	1/4-20 / 30-50	#12-24 / 30-50	1/4-20 / 60-80
Cylinder to Crankcase Nuts	5/16-24 / 150-180				
Stator Plate Screws	1/4-20 / 60-80	1/4-20 / 60-80	1/4-20 / 60-80	1/4-20 / 60-80	1/4-20 / 60-80
Fan Housing Screws	1/4-20 / 60-80	1/4-20 / 60-80	1/4-20 / 60-80	1/4-20 / 60-80	1/4-20 / 60-80
Gear Box Cover Screws	1/4-20 / 60-80	1/4-20 5/16-18 / 60-90 80-100			1/4-20 / 60-80
Engine Mtg. Bolts	5/16-18 / 120-150	5/16-18 / 120-150	5/16-18 / 120-150	5/16-18 / 120-150	5/16-18 / 120-150
Starter Mtg. Screws	1/4-28 / 30-40	#10-32 / 30-40	1/4-28 / 30-40	#10-32 / 30-40	1/4-28 / 30-40
Muffler Mtg. Nuts and Bolts Note: Rotary Engines	5/16-18 / 60-80	5/16-18 / 60-80	5/16-18 / 60-80	5/16-18 / 60-80	1/4-20 / 60-80
Exhaust Manifold Muffler Mtg. Nuts and Bolts Note: Reel Mower Engines	15/16-29 / 80-100	5/16-18-29 / 80-100	5/16-18 / 80-100	5/16-18 / 80-100	5/16-18 / 80-100
Flywheel Nut	7/16-20 / 300-360	7/16-20 / 300-360	7/16-20 / 300-360	7/16-20 / 300-360	7/16-20 / 300-360

Figure 153
Short Block Components for the Model 321 Engine..

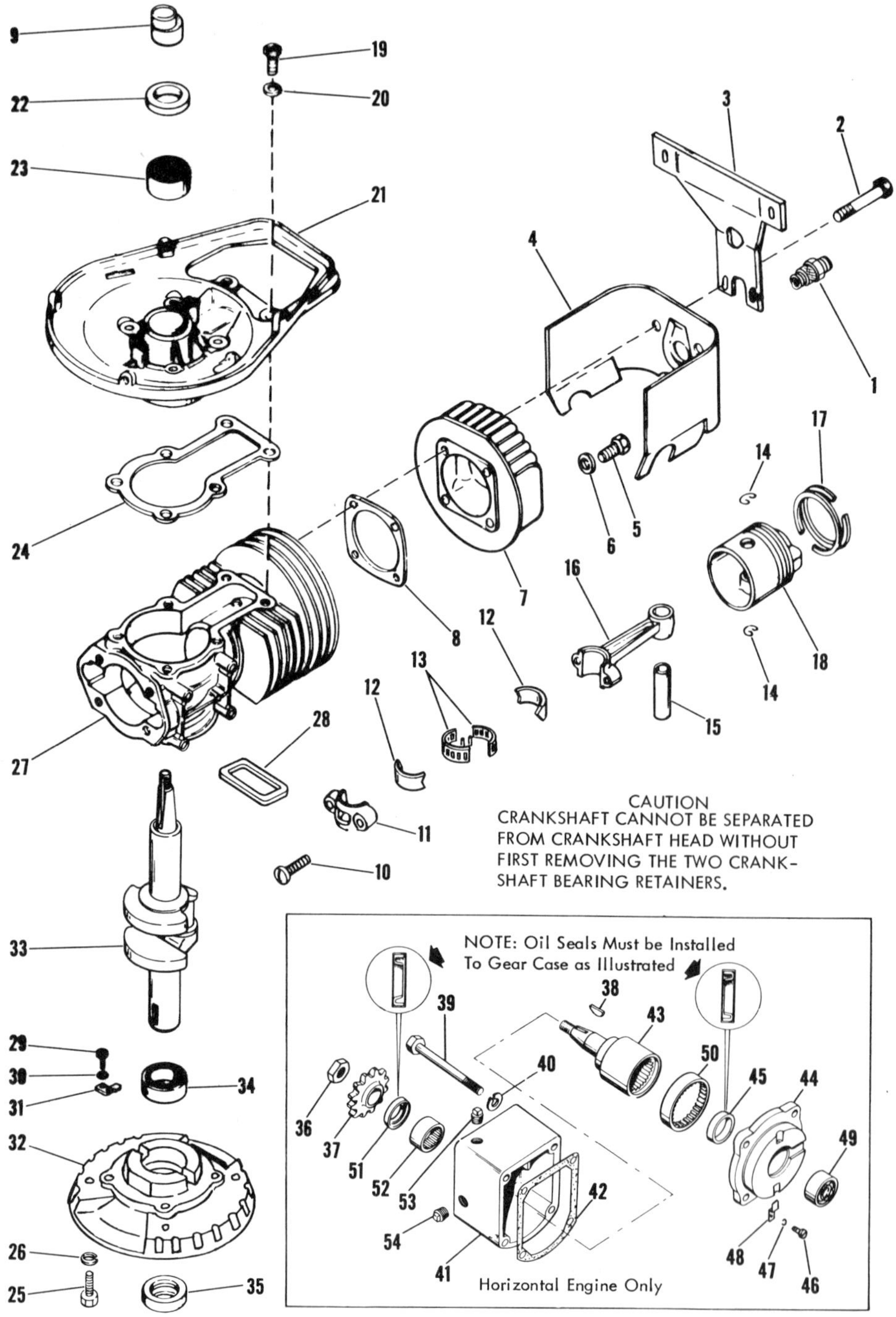

Major Overhaul.

The major overhaul is not an amateur undertaking. If you have more than a basic knowledge of mechanics, or a complete understanding of principles related to internal combustion engine operation, then by all means, proceed. If not, don't try it.

1. Dismantle the engine in sub-assembly units wherever possible.
2. Engines in need of major overhaul are generally corroded, rusty, and many parts are "frozen". Use penetrating oil as needed, but never force parts off. Don't pry components apart unless care is exercised.
3. Always place parts in order of removal. It makes reassembly much easier. If you're uncertain, use the number system.
4. Review figures 146 & 153 for sequence of removal by number.
5. Follow steps involved in magneto removal then continue to crankcase, cylinder, piston, crankshaft and related components.
6. Remove muffler assembly or exhaust manifold if mounted directly to cylinder. Remove any baffle plates still attached to the head.
7. Remove four cylinder head nuts and plain washers.
8. Remove cylinder head and gasket. See figure 154.
9. If cylinder can be removed from crankcase, remove four nuts holding cylinder to crankcase and note position of exhaust port opening.
10. Move piston to bottom dead center and tap cylinder at the base with a soft head mallet to separate from crankcase. See figure 155.
    a. Don't use screwdriver or similiar tool to pry parts free.
11. Unlock connecting rod cap screws, remove cap and screws.
12. Remove piston and rod assembly.
13. Remove wrist pin, after removing retaining rings, and pull the rod away from piston.
14. When working with piston, never place it in a vise.
    a. Never use a tool which will mar or damage ring grooves since compression is dependent on condition of bottom face of ring groove.
15. Remove three cap screws from engine base and press crankshaft out of crankcase. If press is not available, use a rubber mallet.
    a. Remove crankshaft bearing retainers before removing crankshaft from crankcase head.
16. Remove crankshaft from engine base with rubber mallet.
17. Don't remove main bearings unless it's necessary. To remove, use a puller.

NOTE: Every tool or special device listed in this chapter can either be purchased from an automotive jobber, or rented from a local rental agency.

18. After all parts have been removed, inspect for damage, wear or anything that will render them unusable.

Figure 154

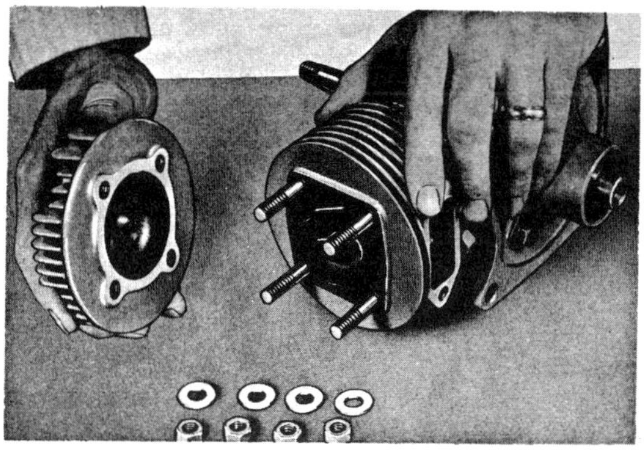

Figure 155

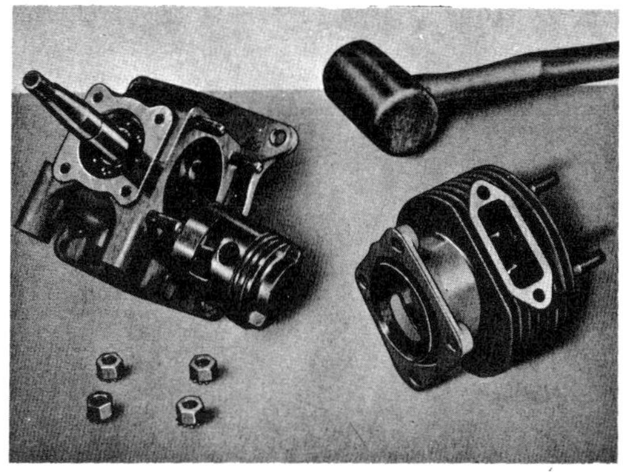

19. Clean all components in a grease solution such as naptha or Bendix cleaner. Remove all carbon deposits and clean all gasket surfaces.
20. Refer to the chart on page 96, and check tolerances on every item.
21. Pay special attention to ring grooves and rings. They must ride free in grooves.
22. When an oversized wrist pin is needed, ream the hole in piston to exact size required. Connecting rod must also be enlarged. When re-installing wrist pin, remember the hole always faces intake port side.
23. The piston is available in .020" oversize from manufacturer. Use a new piston if side clearance of new rings exceed .010."
24. The cylinder can be bored to .020" oversize if walls are badly worn.
    a. If boring isn't required, it's still a good idea to run a deglazer through the cylinder to aid in seating the new rings.
25. Replace the rings on EVERY overhaul job.
26. Clean and check bearings before installing crankshaft.

Figure 156

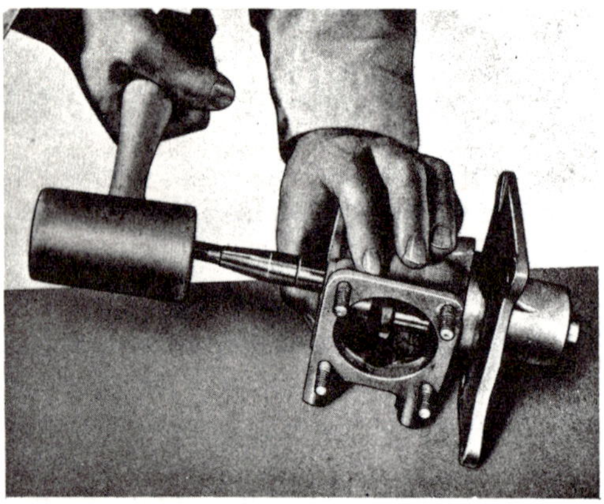

Figure 157

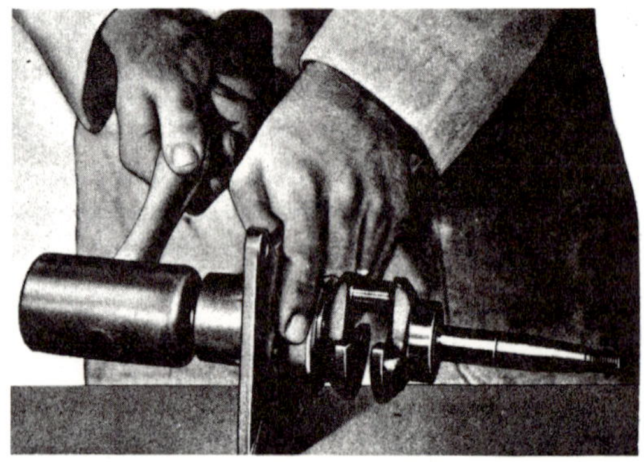

Figure 158

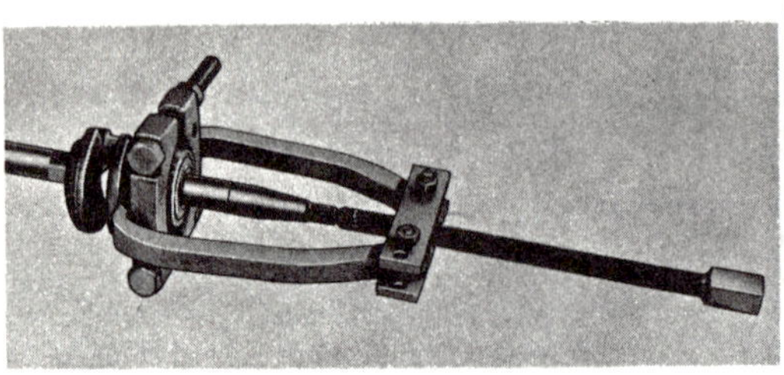

Figure 158a

Reassembly.

1. Refer again to figures 146 & 153. If you followed instructions, and placed parts in order of removal, replacement should be simple.
2. Place new rings on piston with a ring expander. Rings must fit in place freely. Stagger the gaps to reduce compression loss.
3. Assemble piston and connecting rod with wrist pin and retainers.
4. If new crankshaft bearings were installed, be certain crankshaft is supported between counterweights to prevent springing or bending the shaft. See figure 159 & 159a.
5. If engine contains solid rod bearings, assemble the rod and piston to crankshaft before installing crankshaft in crankcase. Refer to chart on tolerances for proper fit. After fit has been determined, remove rod and piston assembly from crankshaft.
   a. Cap must be assembled to rod according to match marks. See figure 160, following page.
6. Assemble engine base and crankshaft to crankshaft. Use a new gasket and seat properly before tightening the three cap screws.
7. Model 321 engines feature a connecting rod bearing kit which is used instead of roller bearing cage and liners. 28 rollers are held together by a wax strip(for easy installation) and form the bearing unit.
8. Refer to figure 161, and make certain there are 28 rollers.
9. Remove the backing paper and wrap bearings and wax strip around the center of the crankpin.
10. Apply grease around the pin on either side of the rollers and slide around in a semi-circular bearing guide at each end of the rollers.
11. Turn crankshaft over and slide remaining two guides in place.
12. Install new liners to connecting rod and cap.
13. Position piston in cylinder until rod liner fits against the rollers, and slip on the rod cap with liner and torque the attaching screws to 65-75 pounds per inch. Refer to torque chart on page 97.
    a. On Model 321 engines equipped with aluminum connecting rod, install rod liners as shown in figure 160. Liner ends must match. Lock rod cap screws by staking as illustrated in figure 160.
14. Install cylinder head and new gaskets to cylinder. Place nuts and washers on studs and tighten nuts to torque of 150-180 pounds per inch.
15. Install magneto back plate and stator plate.
16. Install flywheel, housing, cowling, carburetor and air cleaner.
17. As you reassemble, check all parts for correct operation.
18. Lubricate all parts during reassembly.

Figure 159

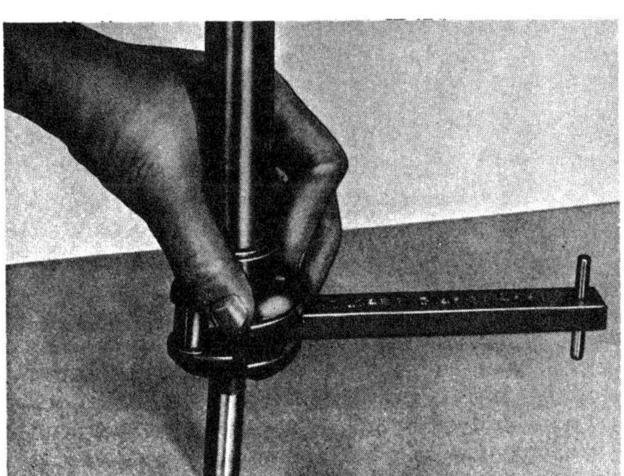

Figure 159a

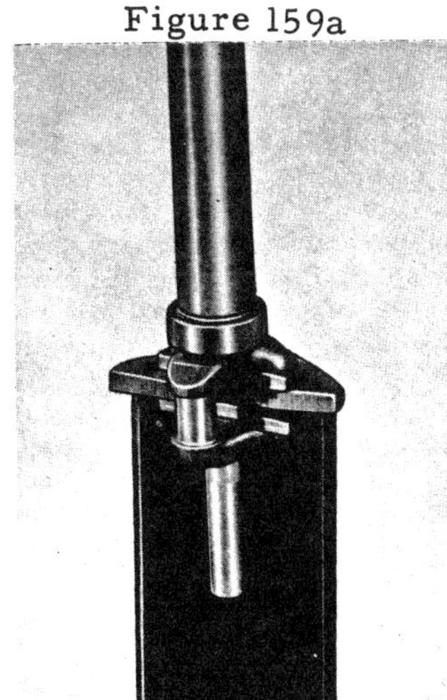

Figure 160

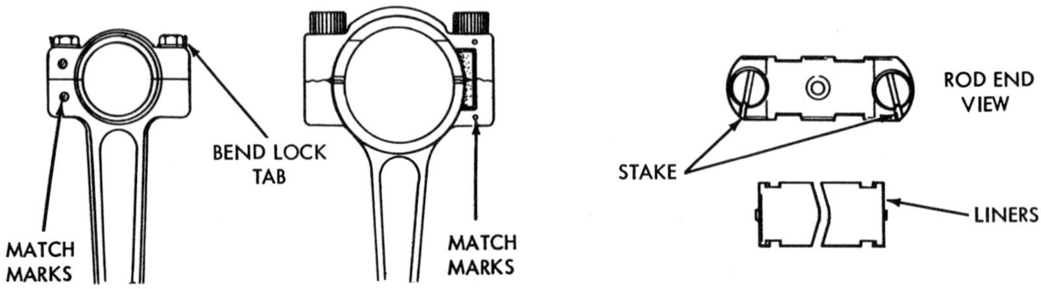

Figure 161

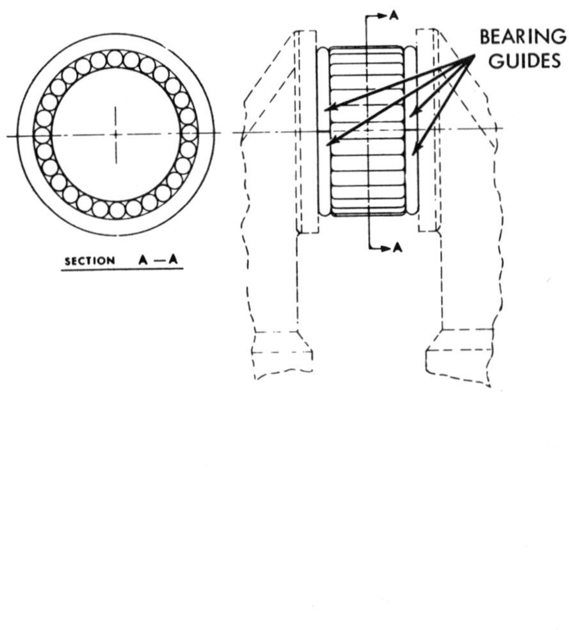

Jacobsen Lawn Mower-18" & 21" Victor Reel Type.

To operate the Victor reel mower, engage both drive and reel controls. See figure 162 & 162a. The drive grip places both controls in the operators hands at all times. Push down front of drive grip(or forward), and power drive to mower is engaged. Pull back(or up) on drive grip and both drive and reel are disengaged, stopping all forward motion and operation.

Engine controls are located down the handle within easy operating range of the user. Controls are similiar to several previously covered.

Figure 162

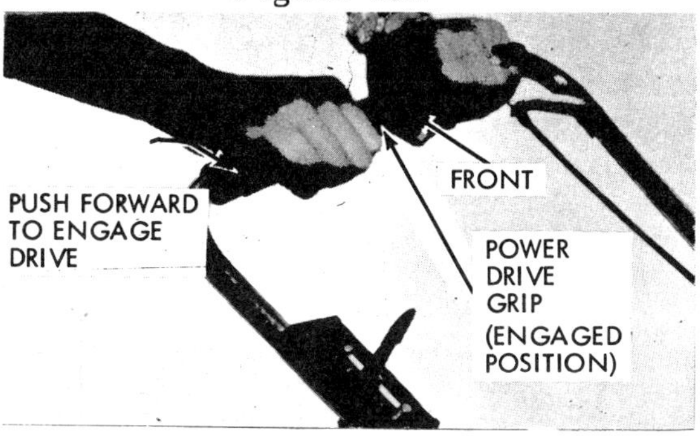

Figure 162a

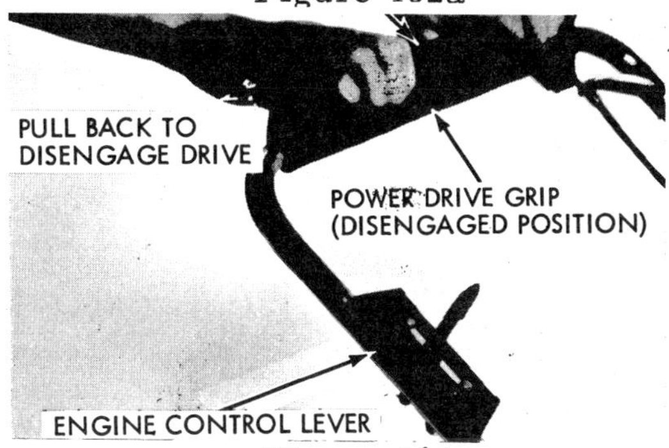

Figure 163

Figure 164

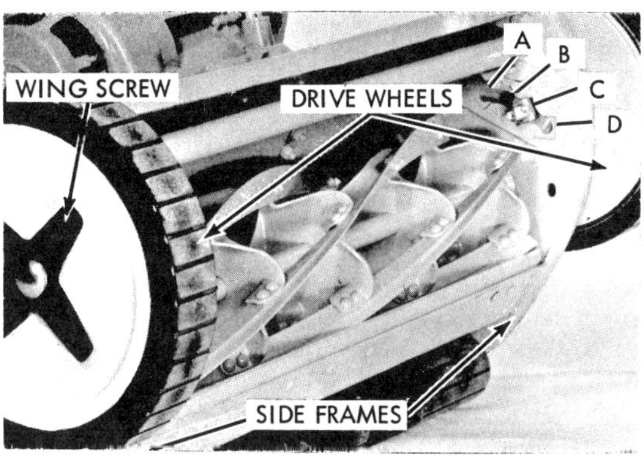

Figure 163a

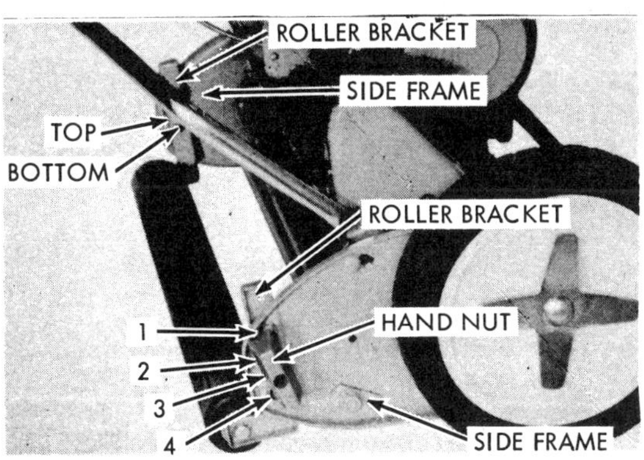

CUTTING HEIGHT	WHEEL POSITION FRONT	ROLLER POSITION SIDE FRAME	NOTCH IN ROLLER BRKT
1/2"	A	1	Bottom
3/4"	A	2	Bottom
7/8"	A	3	Bottom
1"	B	1	Top
1-1/8"	A	3	Top
1-1/4"	B	2	Bottom
1-3/8"	B	3	Bottom
1-1/2"	B	3	Top
1-5/8"	C	2	Bottom
1-3/4"	C	2	Top
1-7/8"	C	3	Bottom
2"	C	3	Top
2-1/8"	D	1	Top
2-1/4"	C	4	Bottom
2-3/8"	C	4	Top
2-1/2	D	3	Bottom
2-5/8"	D	3	Top
2-3/4"	D	4	Bottom
3"	D	4	Top

Cutting heights over 1-1/2" should be used only when grass conditions permit, i.e., tall, firm, vigorous grass.

Cutting Height Adjustment.

1. The cutting height is adjusted by raising or lowering drive wheels & roller brackets. See figures 163 & 163a previous page.
2. Refer to figure 164 and set wheels and rollers to desired height.
3. Loosen drive wheel wing screws and push screws inward, to disengage the special nuts from adjustment notch in side frames.
4. Turn wing screw in the left wheel clockwise.
5. Turn wing screw in the right wheel counterclockwise.
6. Set wheels in desired position and engage two corners of nuts in the notch in the side frames.
7. Tighten wing screws securely to lock wheels in position.
   a. Be sure drive wheels are in the same notch in each side.

Rear Roller Bracket Adjustment.

1. Loosen roller bracket hand nuts until end of nut clears side frame. See figure 163a, page 103.
2. Move roller brackets to desired setting and set nuts in holes in the side frame. (Holes 1, 2, 3 or 4, as desired).
3. Move roller brackets until bolts align with top or bottom notch.
4. Tighten hand nuts after placing square shoulder of bolts in the notch in the roller brackets.
   a. Bolts must be placed in same notch of roller brackets.

Handle Height Adjustment.

The handle height can be adjusted to one of four different positions.

1. To adjust, remove hex nut and handle stop bracket from the handle stop bolt in the left side frame.
2. Remove hex nut and lockwasher from both sides and remove bolts.
3. Push handle forward to desired height and insert stop bolts.
4. Reassemble in reverse order.
5. When handle is in high position, no stop is required.

Figure 165

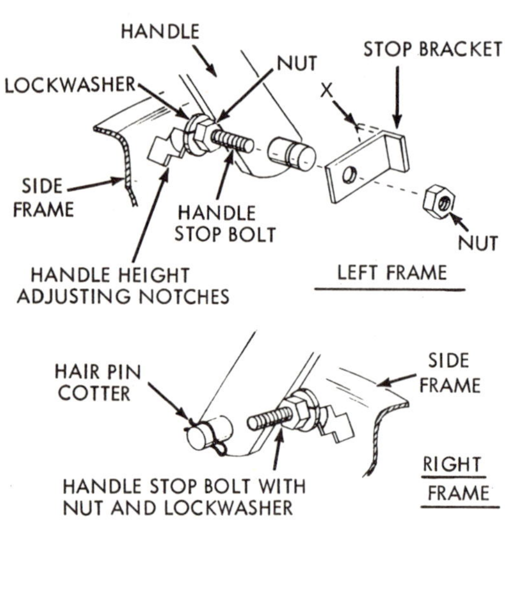

Figure 165a.

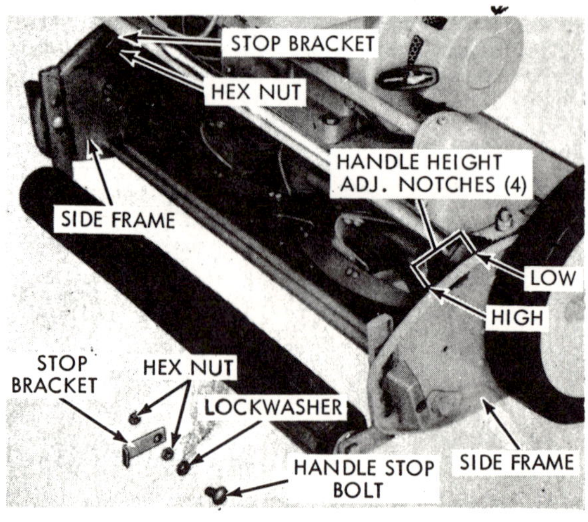

Drive Belt Adjustment.

The spring loaded idler pulley on the clutch automatically compensates for normal belt wear and stretch. If slippage occurs, and belt is in good condition, adjustment is required. See Figure 166

1. Loosen the four engine mounting screws.
2. Slide engine forward to remove excess slack from the belt.
   a. Do not slide engine too far forward or belt will not disengage.
3. Hold engine in place and tighten engine mounting screws.
4. Check pulley alignment, both engine and large pulley must align.
5. Check belt adjustment before proceeding.
   a. Place drive grip in disengaged position.
   b. Move engine control lever to STOP position.
   c. Pull starter handle several times. If mower moves, the belt is too tight. Move engine back slightly to loosen.
   d. Start engine and check clutch. Mower shouldn't move when the clutch is disengaged.
6. To replace the belt, remove belt shield.
7. Loosen engine mount screws and move engine back to extreme.
8. Place drive grip in disengaged position.
9. Remove old drive belt from around engine & large pulley.
10. Place new belt around the large pulley.
11. Slip the belt around engine pulley with bottom of belt between idler pulley and the belt guide. See figure 166a.
12. Replace belt guard with shield pin above the belt.
13. Adjust per above instructions.

Cutter Bar Adjustment.

1. Adjust cutter bar by turning two adjusting screws on inside of each side frame. See figure 167, following page.
2. Turn each screw equally counterclockwise to bring cutter bar closer to the reel blades.
3. The cutter bar should just lightly touch reel blades.
4. Use the newspaper test referenced previously.
   a. Reel should cut newspaper placed on cutter blade and still turn freely.

Figure 166

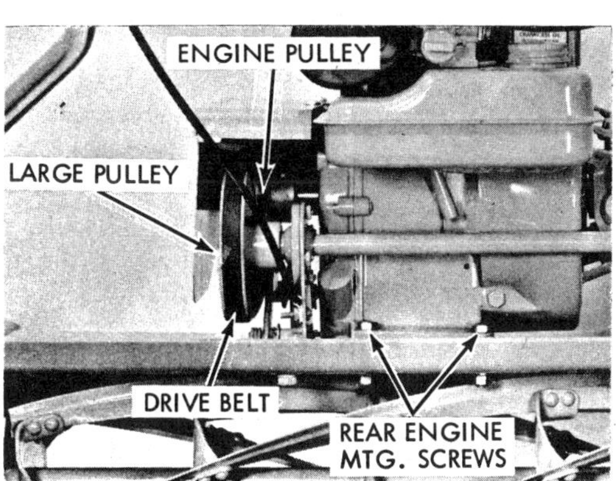

Figure 166a

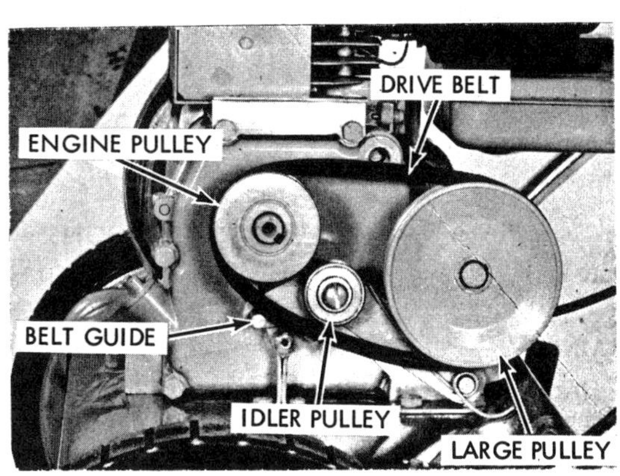

Figure 167

Figure 168

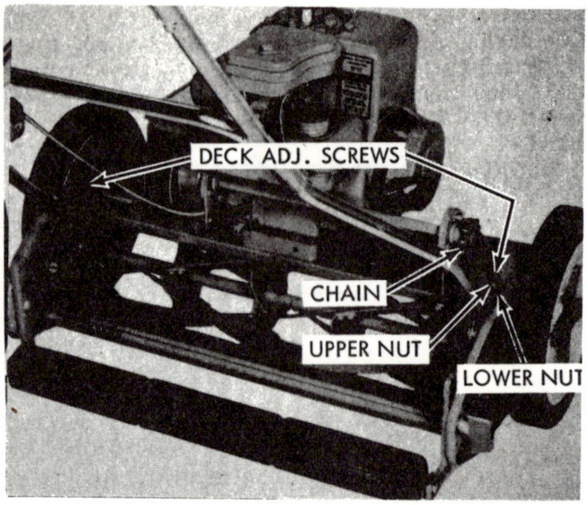

Chain Adjustment.

After the mower, when new, has been used for 5 or 10 hours, check the chain. If there is excess slack, adjustment is required.

1. Check the chain each time it is lubricated.
2. Place drive grip in disengage position.
3. Remove the chain shield.
4. Loosen the upper nut on deck adjusting screws. Figure 168
5. On the chain side of deck, insert a screwdriver or steel bar between the deck and side frame.
6. Pry the deck upward to tighten the chain.
   a. Check the chain for slight amount of slack.
7. When proper tension is attained, reach through handle slot opening in the deck and turn lower nut on the adjusting screw up until it contacts the underside of the deck. Tighten upper nut securely against the deck.
8. Adjust the opposite of deck in the same manner.
9. Reinstall the chain shield.
10. To prevent excess wear, don't tighten chain too tightly.

Lubrication.

Figure 169

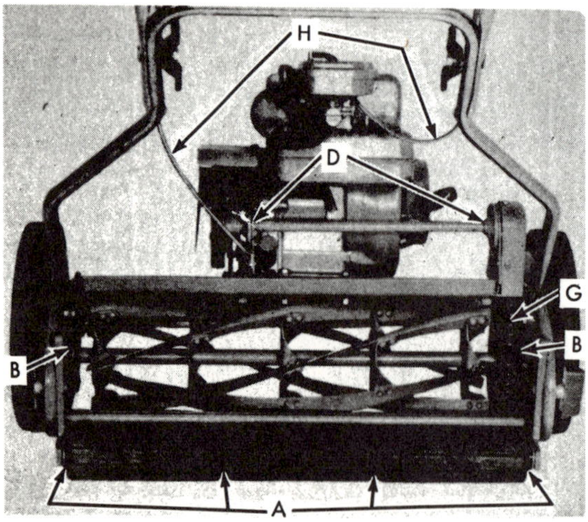

Figure 170

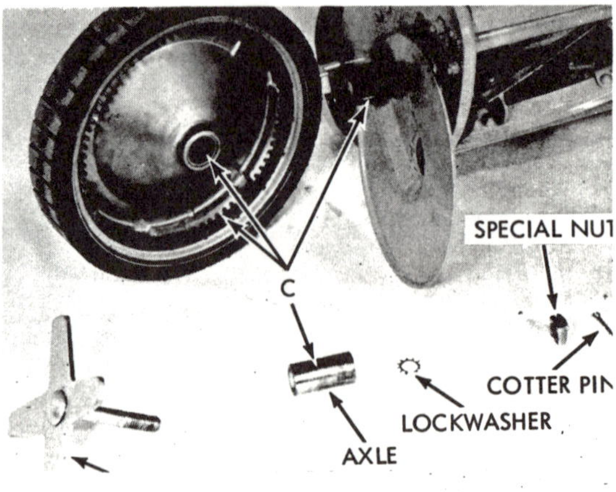

1. Lubricate the rear rollers every 10 to 15 operating hours. Use the SAE 30 and tip mower on side to allow oil to run into rollers. Fig. 169.
2. Reel bearings should be checked when reel blade is sharpened.
3. Lubricate gears and wheels at the start of season. Use a semi-fluid grease such as Alemite Lubricant 0-164. See figure 170.
    a. Remove cotter pins from end of wing screws.
    b. Turn wing screws out of special nuts and remove wheels.
    c. Remove one wheel at a time since each side is threaded differently. Turn wing screw out clockwise on left wheel.
    d. Apply grease to pinion and wheel gears, hubs and the axle.
    e. Reassemble in reverse order.
4. To lubricate cross shaft bearings, use SAE 30 in an oil can and insert through holes of bearing covers. Saturate the felt retainers.
5. Oil clutch pivot points at the beginning of each season.
6. Lubricate the idler arm cam at the beginning of each season, with a light application of grease along the bottom edge of the cam.
7. Oil the chain frequently with application to chain which reaches the inside of each roller.
8. Oil throttle cand clutch control cables. Insert in ends.
9. Change engine crankcase oil per instructions on engine service.

Jacobsen 4-Blade Rotary Mower.

Figure 171

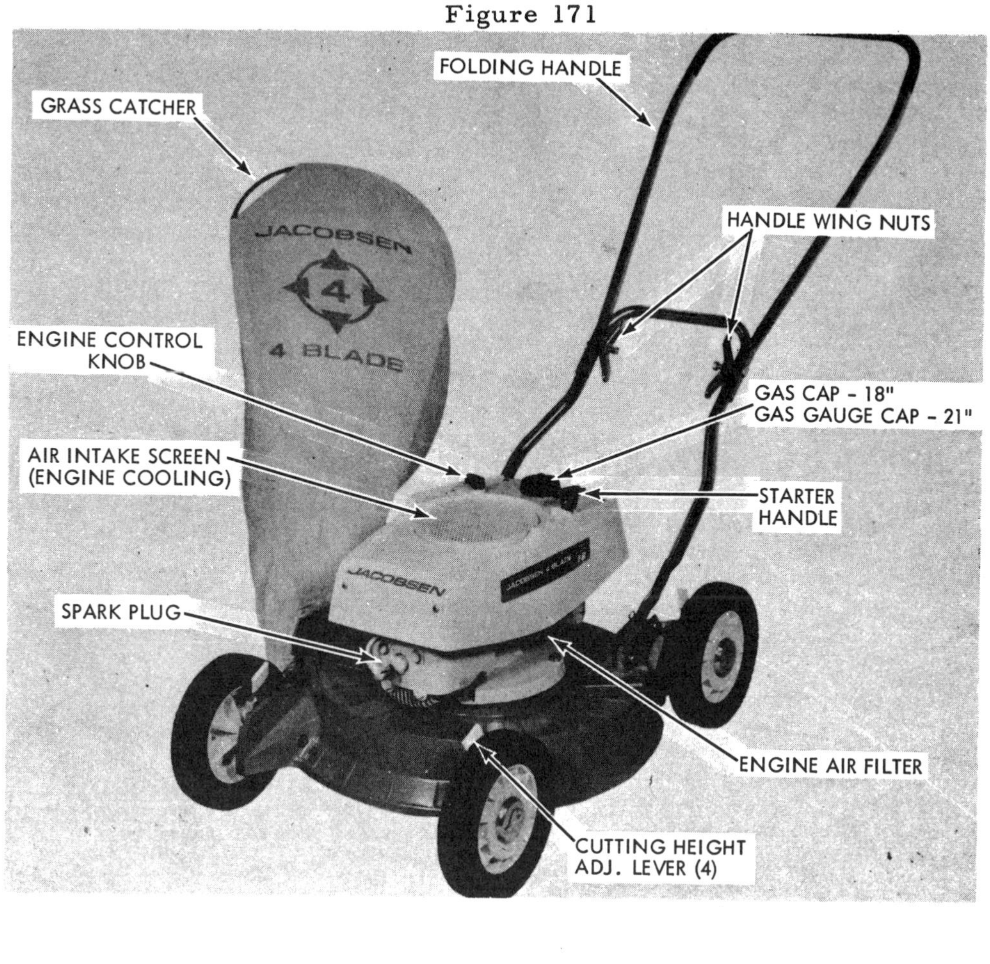

Adjusting the Cutting Height.

1. Always stop engine before adjusting cutting height.
2. Check figure 172, and determine cutting height desired.
3. Move indexing finger against wheel.
4. Move indexing lever and wheel to selected mark and release lever.
   a. Pin in lever must engage hole in quadrant.
5. Adjust all wheels to same height.

Figure 172

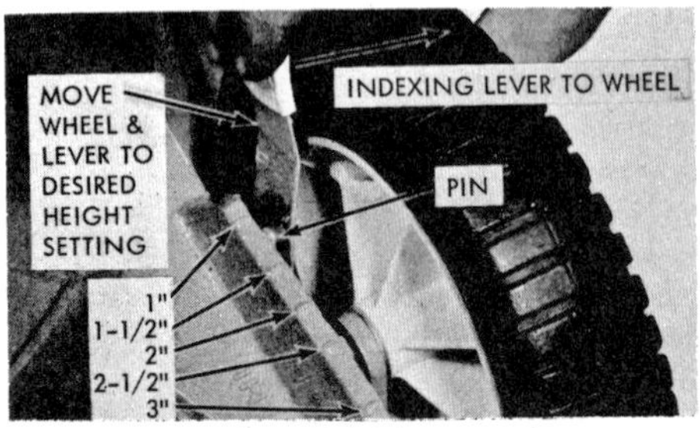

Figure 172a

Figure 173

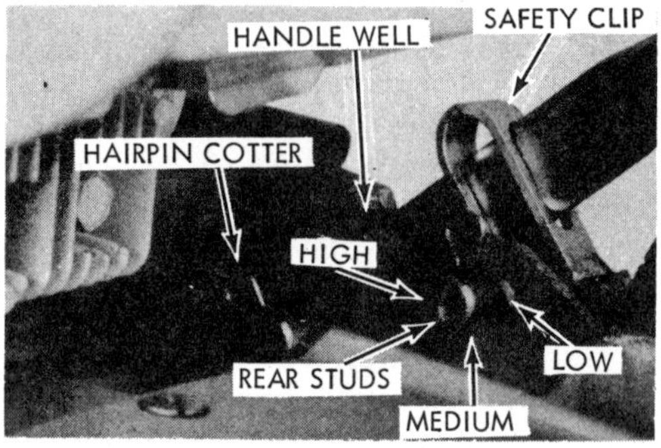

Adjusting Handle Height.

1. To adjust, remove hairpin cotter pins from rear studs in the handle well and remove studs.
2. Relocate studs at level of handle height desired.
3. Replace cotter pins in studs. Studs must be in same position in both handle wells. Safety clip must be hooked behind right rear stud.

Cutter Blade Service.

1. For best cutting results keep the blade sharp.
2. The cutter blades are reversible. When original edge becomes too dull to perform well, reverse the part to obtain a new edge.
3. After both edges have been well used, remove and sharpen.
4. Use a good grinding wheel and follow original angle of blade.
   a. Sharpen each blade equally to maintain balance.
   b. If mower vibrates after sharpening blades, recheck all blades and be certain they were ground down equally.

Figure 174

Figure 174a

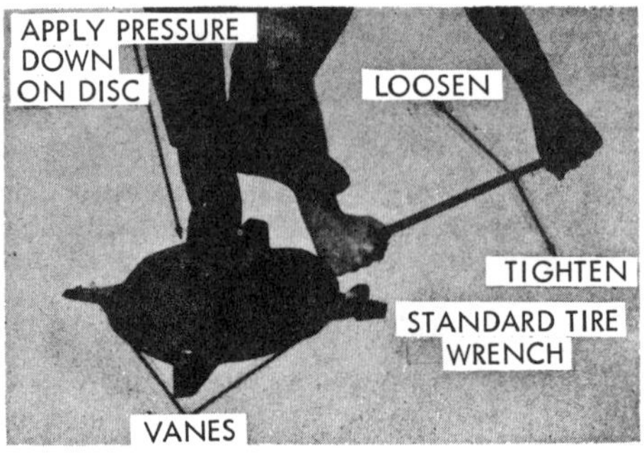

5. If necessary to remove the cutter disc, manufacturer suggests you obtain a special disc lock(part # 112808, Jacobsen), to hold disc in place when removing or reinstalling. See figure 174
   a. If lock is not available, use a wedge but don't damage unit.
6. Remove the disc with a heavy duty wrench(tire wrench will do).
   a. Remove the hex head screw and lockwasher from disc center.
7. To remove blades or vanes from disc, place disc in a vise and use the same heavy duty wrench.
8. If vise is not available, place disc on sidewalk and hold. Fig.174a.
9. If disc was removed to remove only the vanes, remove only the nuts holding vanes to disc. Replace nuts and hold blade in position.
10. To replace or reverse blade, remove vanes, blades and screws.
11. Reverse or install new blades as required.
12. If blade mounting screws are damaged, replace with same type.
13. Refer to figure 175 for proper steps to replace blade.
14. Insert bolt through blade, through disc with flat shoulders on screw engaged in slotted hole in disc.
15. If vanes are utilized, attach them over opposite blades.
16. Nuts must be positioned with hollow side facing the disc.
17. Position each blade with end straight out from center of disc.
   a. Vanes must be off-set to fit over edge of disc. Fig. 175.
18. Tighten the nuts securely.
19. Position disc on adapter and insert mounting screw with lockwasher through center hole.
20. Tighten disc mounting screw with the heavy-duty wrench.
20. If disc lock was utilized, remove it from the deck.

Turned Blade.

1. The blade is designed to turn when hitting a solid object(figure 174b) which prevents serious damage to the crankshaft.
2. Mower will generally vibrate excessively after this occurs.
3. Make an inspection. If mower blade is damaged, replace it.

Figure 174b

Figure 175

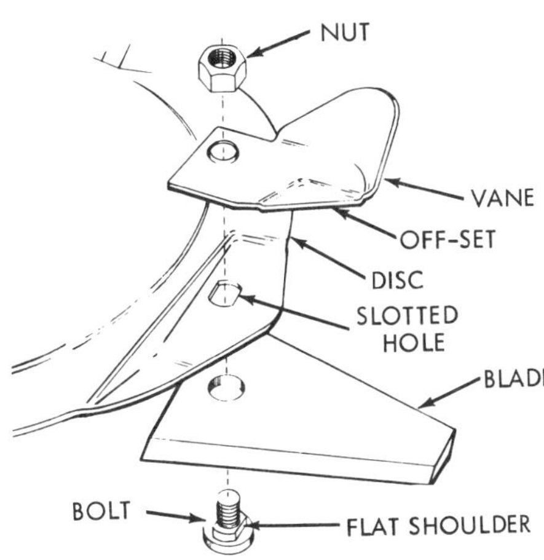

Wheel Care.

1. To remove the wheel, squeeze indexing lever and move beyond adjustment range(forward on front wheels, backward on rear wheels) until nut on the end of axle bolt is exposed.
2. Remove the nut.
3. Straighten the 3 tabs of inner wheel hub cap with a screwdriver and remove from the wheel. See figure 176.
4. Remove the axle from wheel support. Turn counterclockwise.
5. Remove flat washer and axle bolt from wheel.
6. Reassemble in reverse order and make certain flat washer is placed on axle before mounting wheel on wheel support.
7. Tighten axle bolt to wheel support(figure 178).
8. Place nut on axle nut and tighten securely.
9. Insert hub caps in slots and bend tabs to secure.

Lubrication.
1. Lubricate wheels and support shafts once a month with SAE30 oil.
2. Spin wheel to allow oil to penetrate into bearing.
3. Oil the axle and hub at back of wheel.
4. Oil wheel support shaft where it enters the deck.
   a. Move index lever back and forth to work oil into bearings.
5. In extremely dusty areas, bearings become tight and require cleaning. Remove wheel per previous instructions and clean axle-bearings.

Figure 176

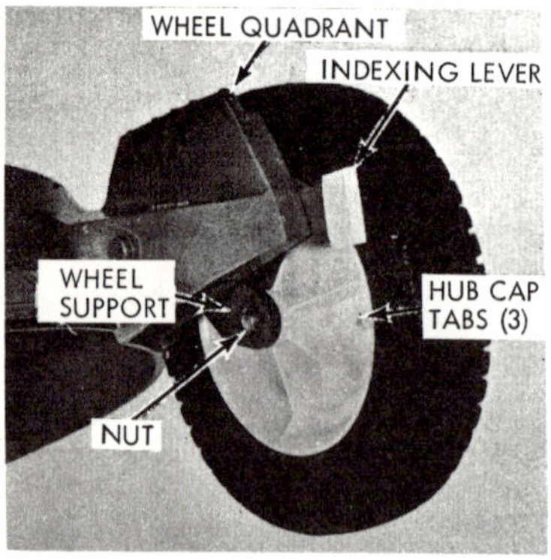

Figure 178

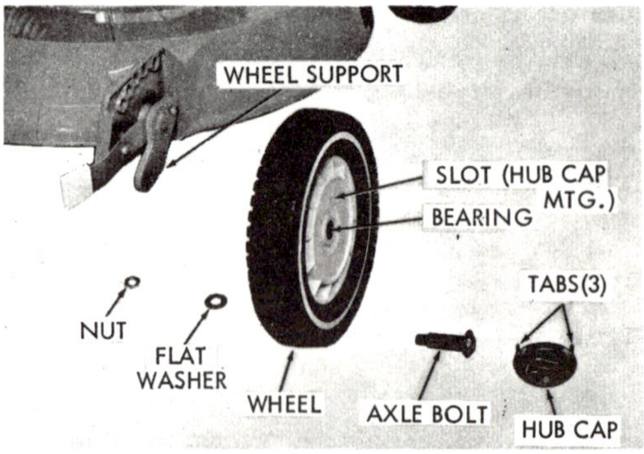

Figure 177

Figure 179

The Jacobsen "21 Manor Mower".

Traction Clutch.

1. Refer to figure 180 for location of traction control lever. The control lever engages the traction clutch to set mower in motion.
2. The lever has two operating positions.
   a. Forward, away from the operator, engages the clutch.
   b. Back, toward the operator, disengages the clutch.
3. To adjust the clutch rod length, refer to figure 181.
4. Clutch rod adjustment is necessary when forward position of the lever doesn't engage the traction clutch and move the mower.
   a. Rod adjustment must be changed if handle height is altered.
5. Remove anti-rattle spring.
6. Remove adjustable coupling from lever.
7. Loosen locking nut on the coupling.
8. Adjust the coupler as needed.
9. Tighten locking nut and replace.
10. Don't force lever forward to engage the clutch.
    a. If clutch doesn't engage easily, refer to above steps.
    b. Forcing the lever could bend the rod and distort the action.
    c. If rod has been bent, remove from unit and straighten.

Figure 180

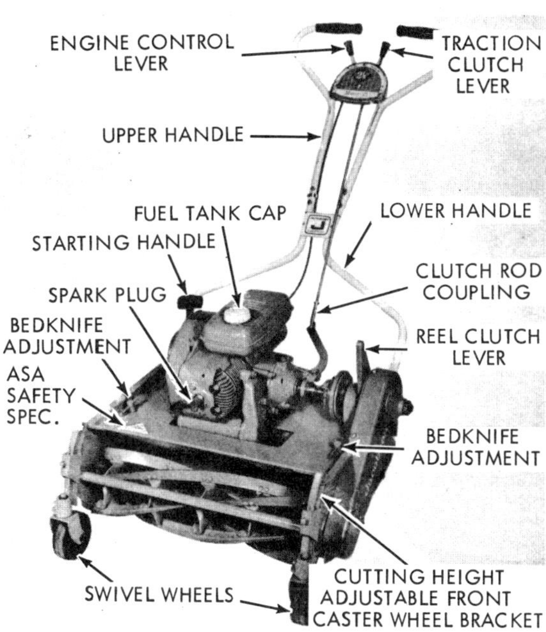

Figure 180a

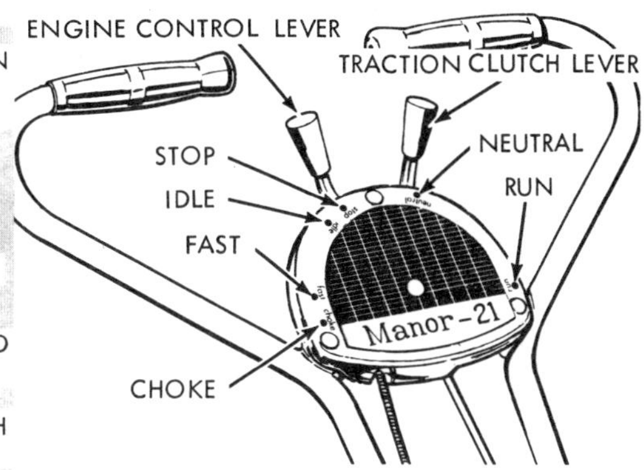

Figure 181

Traction Chain Adjustment.

1. If chain adjustment is required, block unit off the ground. Fig. 182.
2. Loosen 4 screws that hold traction bearing housing to frame.
3. Move rollers downward until slack is removed. Don't overtighten.
4. Tighten screws and hold rollers to retain positioning.
   a. Tighten and position right and left bearing bracket equally.
   b. When all screws are tight, recheck for proper alignment.

Traction Clutch Adjustment.

1. If clutch fails to function and rod position is in adjustment, refer to figure 183, and make clutch adjustment.
2. Loosen fillister head screw in clutch collar.
3. Tap collar toward clutch, very slightly.
4. Tighten fillister head screw.
5. Engage the clutch. It should operate without any slippage.

Reel Drive Belt Adjustment.

1. Refer to figure 180 for location of reel clutch handle.
2. To engage clutch, move to rear. To disengage, push forward.
3. The reel clutch cannot be engaged unless engine is running.
4. If mower is new, check belt after 8 hours of operation.
   a. After break-in period, check belt every 40 hours of operation.
5. Slippage will occur if belt is too loose, or if saturated with oil.
6. Check traction by setting traction clutch lever in neutral.
7. Start the engine.
8. Engage reel clutch and stop the engine.
9. Remove spark plug wire from plug for safety precaution.
10. Engage traction clutch.
11. Place right hand on traction clutch, and left hand on the reel.
12. Pull up and turn reel backward, and hold traction clutch firmly.
    a. If reel turns without turning the clutch, adjust the belt.

Figure 182

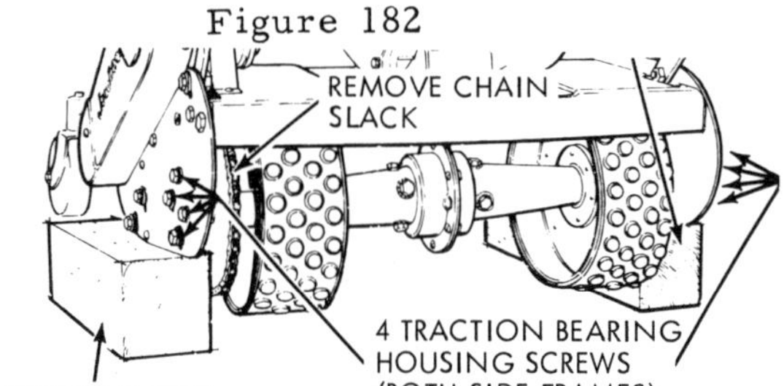

Figure 184

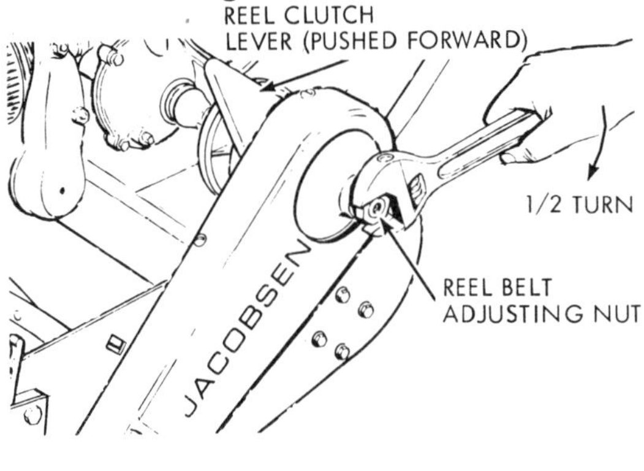

Figure 183

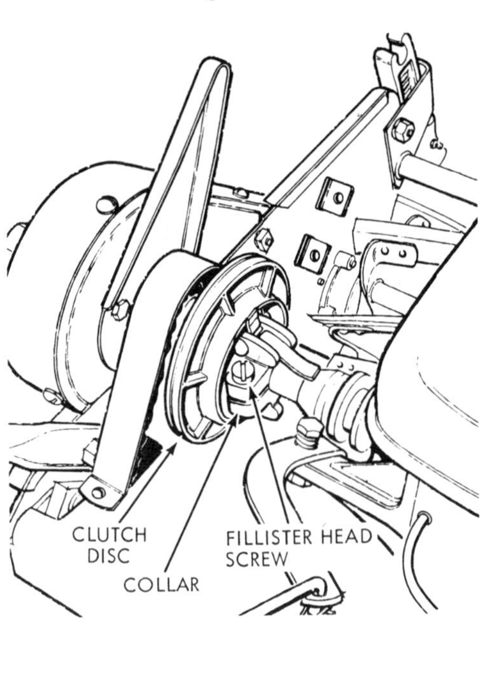

13. Reel clutch must be engaged.
14. Hold reel with left hand and turn adjusting nut(figure 184),clockwise about 1/2 turn. Recheck slippage per previous instruction.
    a. Reel clutch must be disengaged, engine restarted, and the reel clutch re-engaged to position clutch and drive belt.
    b. If the belt still slips, repeat above steps until corrected.
15. Remove the belt shield and check belt condition every 40 operating hours.
16. If belt is oil soaked, clean with cleaning fluid on a rag. Clean only the belt surface. Don't allow cleaner to soak in. Wipe dry before use.
17. Replace the belt shield.

Bed Knife Adjustment.

1. The bed knife requires adjustment when tolerance with reel blade is too wide and the mower cuts unevenly.
2. Disengage reel clutch until reel can be moved freely by hand.
3. A bed knife adjusting nut is located on each side of the mower.
4. To bring bed knife closer to reel, turn nut counterclockwise.
5. If reel is striking bed knife, turn adjusting nut clockwise as needed to correct.
6. Adjust the nuts while turning the reel by hand. When the reel barely touches bed knife, set bed knife adjusting screw at that point.
    a. Use the newspaper test for proper setting.
7. To sharpen cutting edges on temporary basis, set bed knife until it creates a slight binding with reel. Adjust evenly at both ends and work reel at a slow speed for several minutes.
8. Reset for normal operation.
9. Refer to previous text to sharpen permanently.
10. Don't overtighten bed knife setting for normal operation. Both the bed knife and reel could be damaged beyond repair.

Reel Service.

1. Keep the reel free of end play.
2. To adjust, disengage reel clutch.
3. Loosen locking nut on end play screw(figure 185).
4. Turn screw clockwise until end play is removed. Tighten the nut.
5. Spin reel after adjusting, it must spin freely.

Figure 185

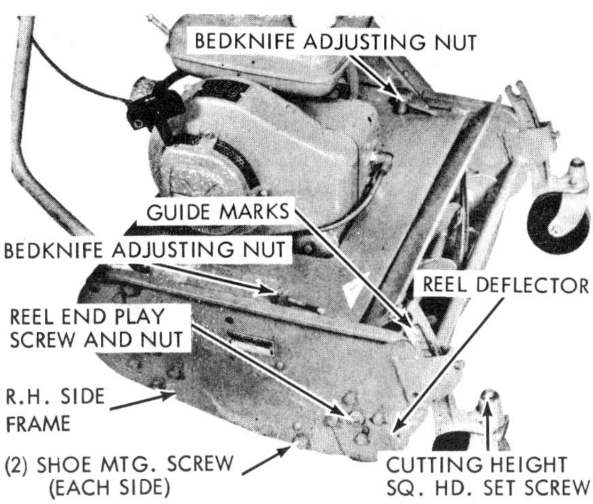

Figure 186

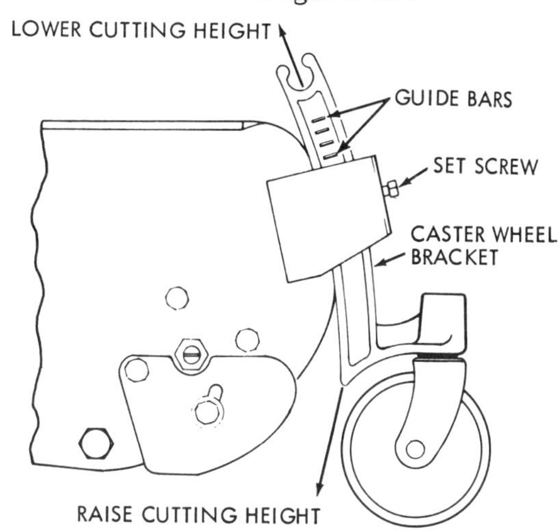

Cutting Height Adjustment.

1. Cutting height may be adjusted from 5/16" to 1 1/2".
2. To adjust, place mower on a level surface, and loosen set screw.
3. Use guide bars and set each wheel to same notch.
4. Move bracket down, cutting height is raised.
5. Move bracket up, cutting height is lowered.
6. Measure distance from floor to top of cutting edge for height.
7. Tighten set screw.

Control Panel Installation.

1. If control panel has been removed, or component replacement is required, proper control assembly is necessary.
2. To install clutch rod, insert long bent end through control lever on control panel and secure with cotter pin.
3. Attach engine control lever components in place on control panel & attach throttle cable to engine.
4. Attach control panel to upper handle.
5. Insert coupling end of clutch rod through hole in traction clutch lever, from the top side. Secure with anti-rattle springs.
6. Check controls for proper setting.
   a. Refer to clutch adjustment section for proper test.
   b. Place engine control lever in choke setting and verify setting at carburetor.

Lubrication.

1. Grease reel bearings(A), figure 188, outer rear roller shaft bearings(B), inner rear roller shaft bearings(C), and reel clutch bearings, (D), with grease gun and semi-fluid grease, every 40 operating hours.
   a. Disengage reel clutch to lube fitting in end of clutch shaft.
   b. Remember, the engine must be running to disengage clutch.
2. Oil traction drive sprocket bearing(E), speed reducer housing on the engine(H) by removing pipe plugs at points H & I, guide wheel bearings (J) which should be lubricated daily, clutch sleeve and fingers(K), caster bracket bearings(L), and the roller chain(every 8 hours), with SAE 30 oil. Daily lubricating except in speed reducer and roller chain will prolong life of mower.

Figure 187

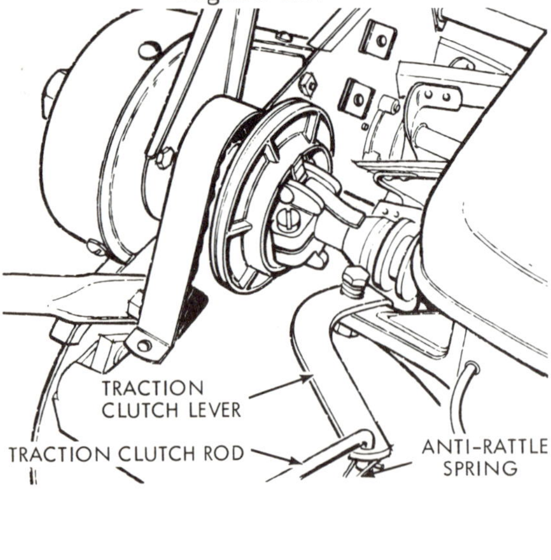

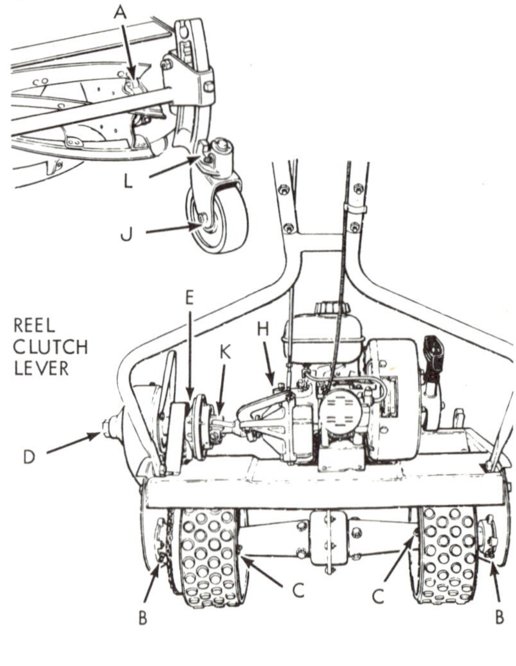

26" Estate Rear Wheel Drive Mower

The Controls.

1. The throttle control lever functions similiar to other units.
2. Traction clutch lever has 3 positions on this model. When the lev-is pushed forward, clutch is engaged and unit moves ahead. When the lever is pulled completely toward operator, it is in reverse. The lev-er must be held in position for reverse movement. Midway between the two settings is the neutral position.
3. Rotation of cutting reel is controlled by Reel Clutch Lever. To en-gage cutting reel, pull lever to the rear

Cutting Height Adjustment.

1. Cutting height settings can be set from 5/16" to 2 1/2".
2. Place mower on a level surface.
3. Loosen both hex nuts which hold guide wheel brackets to frame.
4. Move guide wheel brackets to desired setting. See figure 190.
5. Snug the hex bolts at extreme end of slot and measure height.
6. When setting is made, tighten the screws securely.

Cutting Unit Adjustment. Figure 191, following page.

1. If bed knife requires adjusting, disengage the reel clutch.
2. Remove belt shield & adjustment screw shield, from left frame.
3. Loosen two upper hex head bed knife shoe adjusting screws in frame about 1/4 turn.
4. Tighten the lower screw to raise bed knife closer to reel blade.
5. Rotate reel by hand and adjust lower screws equally until the knife-reel blade assemblies just touch. See previous test with newspaper.
6. Tighten upper screws securely.
7. Replace adjusting screw and belt shields.
8. Refer to previous instruction regarding sharpening bed knife.

Cutting Reel Service.

1. To sharpen reel blades, remove belt shield from left frame.
2. Push reel clutch lever forward and disengage.
3. Use a reversing crank, figure 192, available from Jacobsen.
4. Adjust bed knife evenly(see instructions above) until reel blades are touching bed knife along the entire length.
5. Apply a mixture of oil and carburundum(#60) to reel blades.
6. Turn reel backwards as needed to sharpen evenly.
7. If there are high spots on the bed knife, loosen screw nearest spot

Figure 189

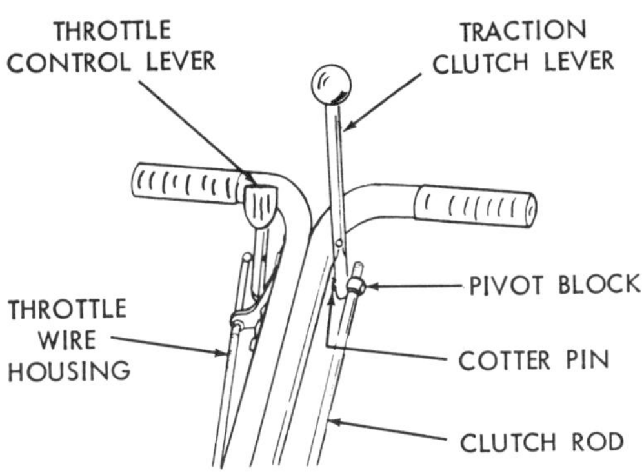

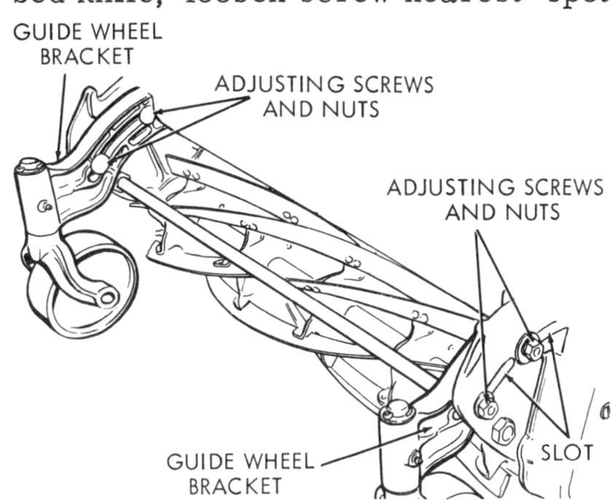

Figure 191

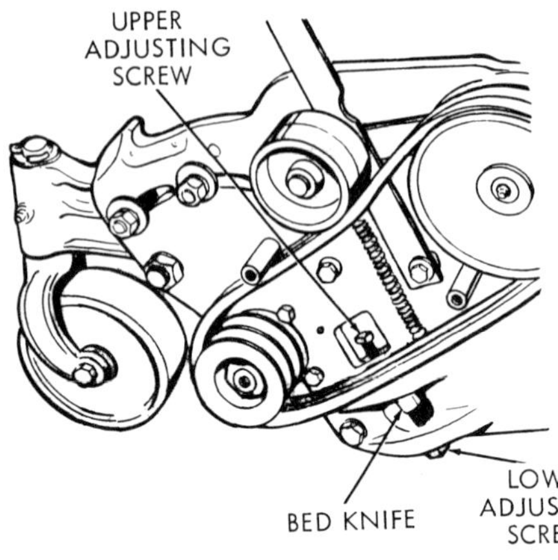

Figure 192

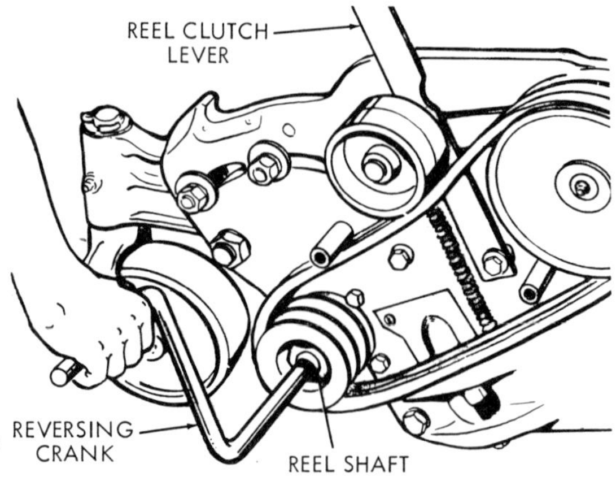

and insert a shim between the bed knife and shoe.

8. When blades have been honed, wash grinding solution away and dry all components. Use a cleaning solvent. Remove crank-replace shield.
9. Reset bed knife and reel setting per previous instructions.
10. To remove reel, remove belt shield, belts, and pulley from reel.
11. Remove bearing cover and housing from side frame.
12. Move away from bed knife and remove from the mower.
13. Replace in reverse order.

Chassis and Drive Adjustment.

1. If clutch slips or fails completely when mower is placed in forward position, loosen the fillister head screw on clutch adjusting collar.
   a. Move collar away from engine slightly, as needed.
   b. Tighten the set screw and test the control.
2. If clutch slips or fails completely in reverse, clutch rod linkage is the part to adjust, by changing position of the pivot block.
3. To adjust, remove cotter pin from pivot block.
4. Turn block along clutch rod(it's threaded) as needed to provide good reverse clutch contact, and keep clutch lever from moving too far back.
   a. Install cotter pin and test. Readjust if required.
5. If reel shaft has too much end play, disengage reel clutch.
6. Move bed knife until there's no contact with reel.
7. Remove cover plate from right bearing housing.
8. Adjust nut on reel shaft clockwise to draw shaft into bearing.
9. DO NOT overadjust. After end play is removed, reel must still be free to turn without binding.
10. Replace all components in reverse order and reset bed knife.
11. See figure 193 for parts breakdown on belts, pulleys and idler.

Reel Clutch and Belt Service.

1. To service reel clutch and belts, remove screws and belt shield.
2. If belt idler spring is broken or weak, remove and replace.
3. If reel belts are badly worn and stretched, replace.
   a. Replace as a matched set.
4. To adjust if not too badly worn, adjust idler to proper position.
5. Replace all components in reverse order.
6. Figure 194 illustrates clutch, shaft and reverse drive components.

Figure 193

Reel Drive Components.

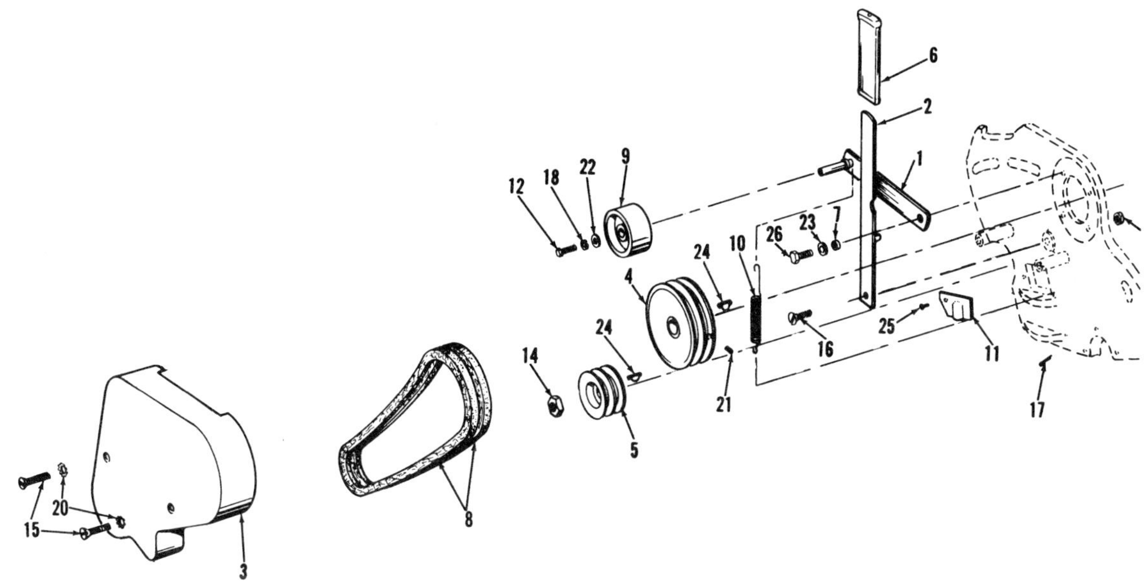

Figure 194

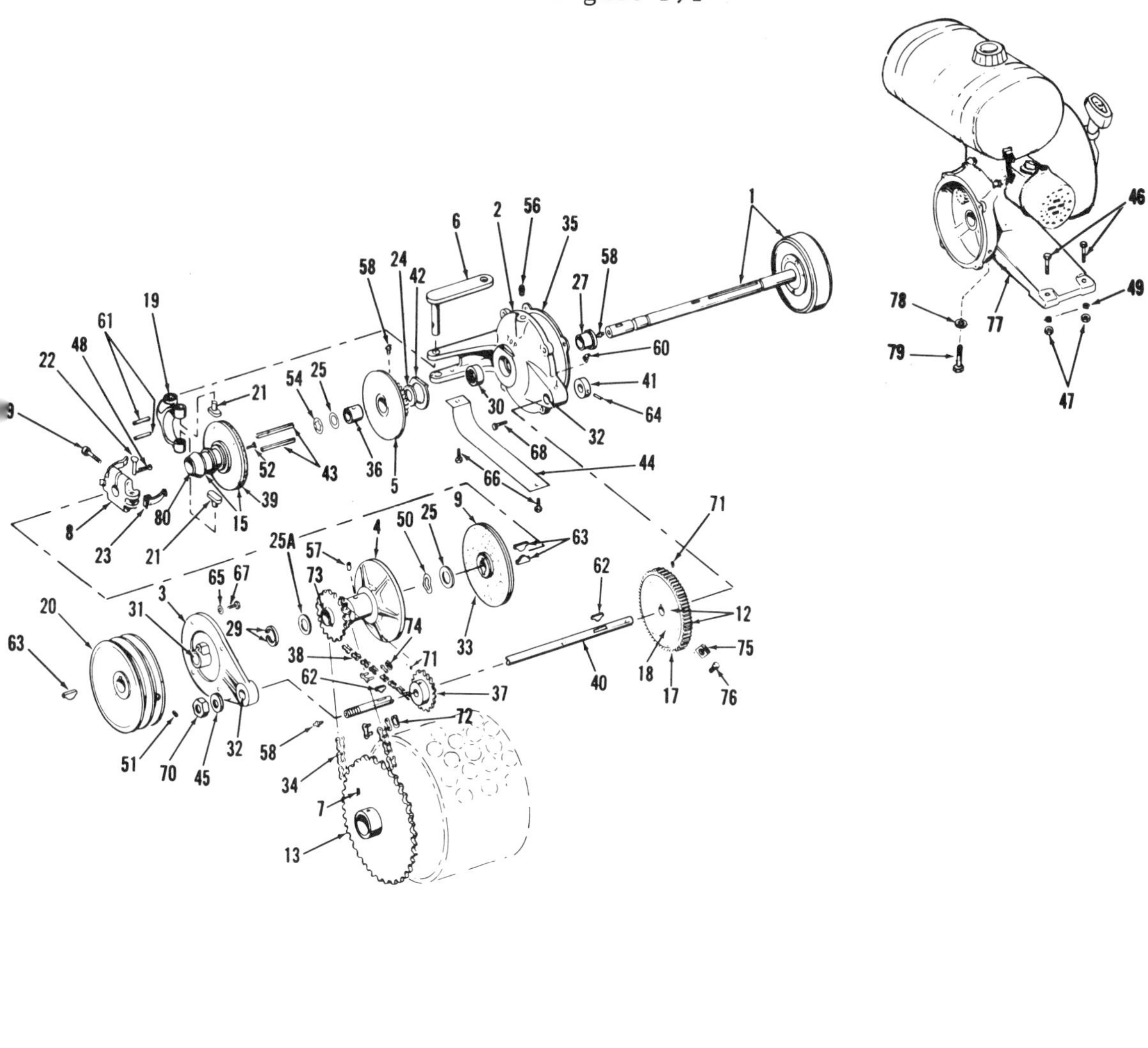

Jacobsen "Javelin Mark II" Riding Rotary Mower

Figure 195

Cutter Height Selection.

1. Change cutting height by adjusting control lever(figure 196), firmly pressing lever down to release pin from the notch and replacing in the notch desired. To raise cutting height, pull back on the lever. Lower cutting height is attained by pushing forward on the lever.
   a. Cutter bar height can be changed whether mower is stopped, or running. High cut is 4" and lower cutting height is 1 1/2".

Cutter Clutch.

1. The cutter clutch lever is located on left deck of mower.
2. To engage, push lever to left and forward until it stops.
3. Disengage clutch by pulling lever to rear and right, against spring tension until it stops.
   a. When clutch is disengaged, a safety brake stops the cutter bar.
4. Always disengage clutch when starting, or moving, the mower.

Figure 196

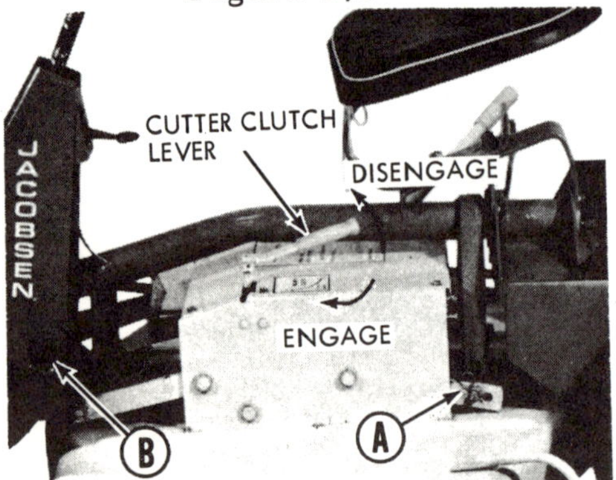

Figure 196a

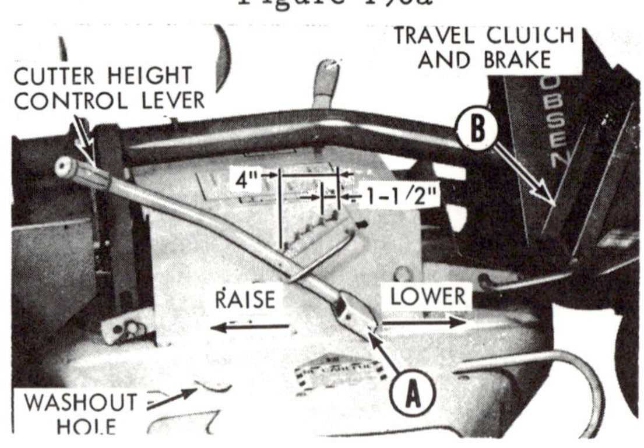

Cutter Bar

1. If cutter bar blades become dull, remove and sharpen.
2. Disengage cutter clutch.
3. Tilt mower up on aft(back) end.
4. Remove cap screw and lockwasher from cutter bar shaft.
   a. Use a 2 x 4 or other wood article and insert through discharge opening to lock cutter bar while loosening the screw.
5. Remove cutter bar and sharpen per previous instructions. Grind each side equally to maintain proper balance.
6. Replace cutter bar with cutting edge facing downward.
   a. Studs on adapter must be fully engaged in two cutter bar holes.
7. Replace lockwasher and cap screw and tighten screw securely.

Cutter Brake.

1. The cutter brake is automatically applied whenever clutch lever is disengaged.
2. Replace the brake cutter pad when replacing cutter drive belt.

Cutter Drive Belt.

1. Check belt occasionally. If it is worn or stretched, replace it.
2. To determine condition of the belt, remove lift bracket cover.
3. Measure belt configuration as in figure 199. If distance shown less than 1", adjust by moving entire cutter housing forward. Figure 200.
4. At the factory, front and rear cutter housing mountings are set, using holes A-1 and A-2. For greater belt tension, first place the clutch lever in OFF position.
5. Remove front & rear mounting shafts, slide housing following, and replace mounting shafts in holes B-1 and B-2.
6. When belt is replaced new, remount housing in holes A-1 & A-2.

Figure 197

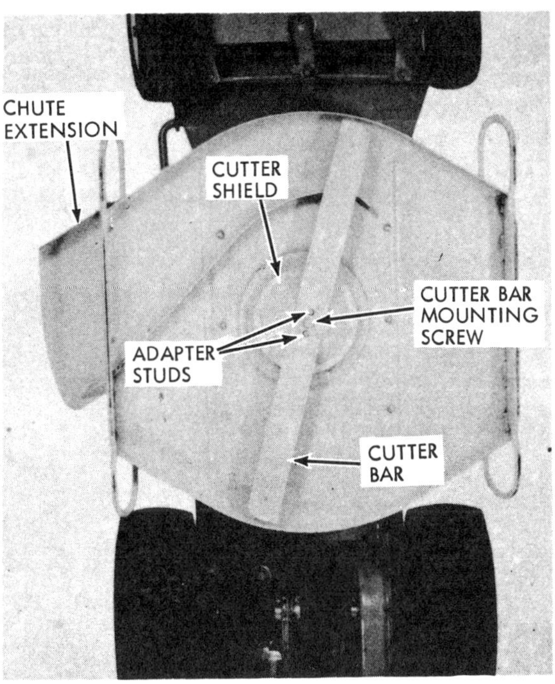

Figure 198

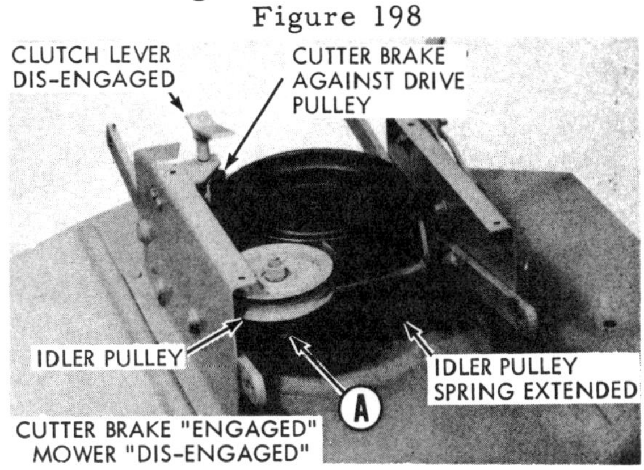

Figure 198a

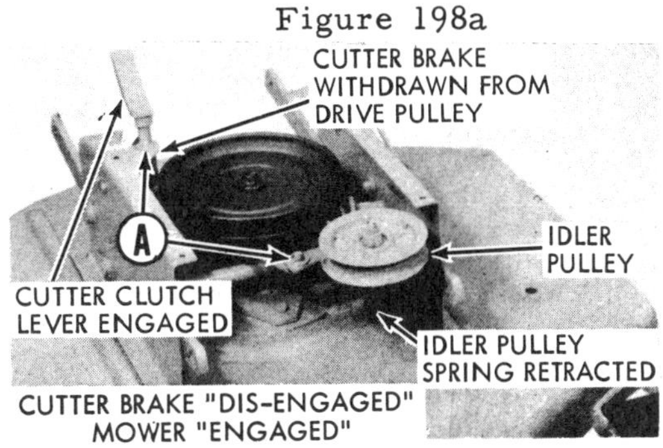

Figure 199

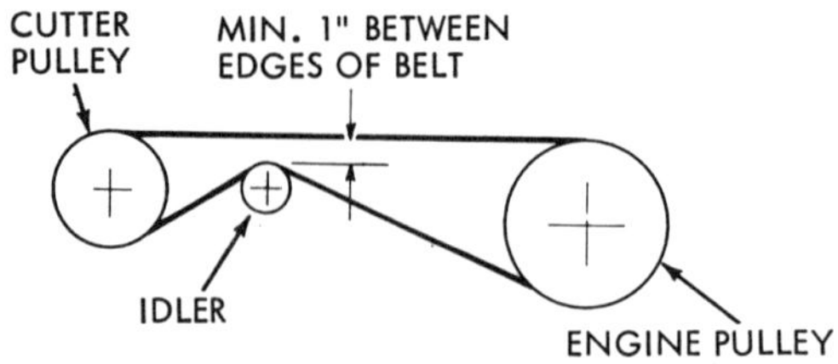

Figure 200

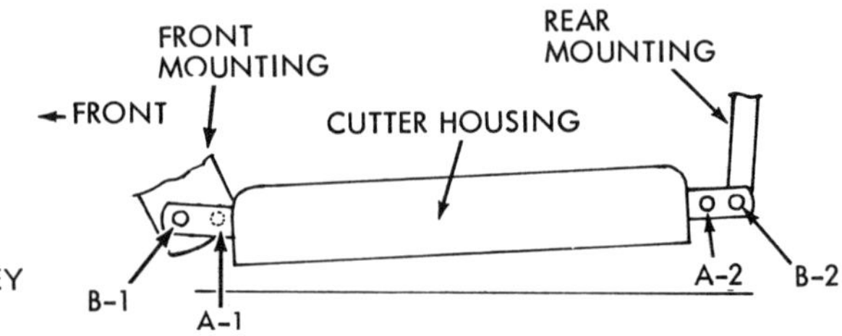

7. Before replacing belt, disconnect the spark plug wire.
8. Remove lift bracket cover by removing six screws with 1/4 wrench.
9. Disengage cutter clutch. Figure 196a.
10. Remove lift bracket cover.
11. Remove belt from cutter pulley(fig. 201a), by moving cutter brake away from pulley and slipping belt past the belt retainer.
12. Stand the mower up on aft(back) end.
13. Push belt down through belt guide and around engine drive pulley.
14. Pull lower end of shuttling gear case(Chain Case Shaft, figure 202) away from engine disc and slip belt between friction drive wheel and the engine disc.
15. Reverse above steps and install a new belt.
    a. To insure proper alignment, engage cutter clutch lever before reinstalling cutter lift bracket cover.
    b. There should be no less than 1/16", or more than 3/16", clearance between belt retainer and rim of idler pulley.
    c. Make certain edge of retainer is not turned into the belt.

Removing Cutter Housing.

1. Follow above steps through belt removal.
2. Remove hairpin cotter pins from front & rear mounting shafts.
3. Drive shaft out opposite side. Use a rubber mallet. Pull the shafts out and remove housing from the mower.
4. Reverse above steps to replace.

Figure 201

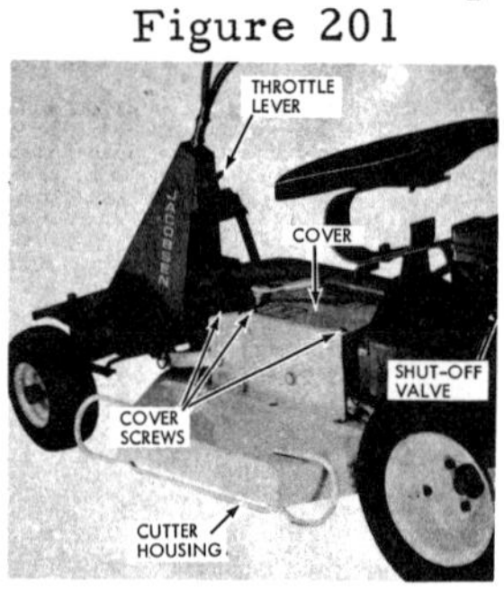

Figure 201a

Figure 202

Figure 203

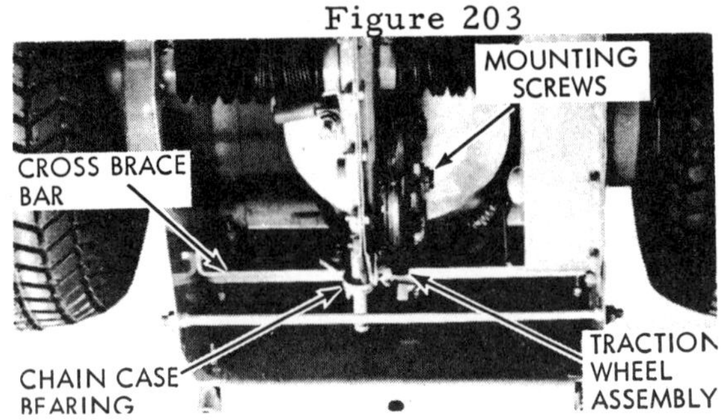

Traction Drive.

1. If slippage occurs, first check drive wheel for wear.
   a. To replace, separate rubber edged drive wheel from wheel back plate. Remove the five screws and lockwashers. Fig. 203
   b. Assemble replacement drive wheel to plate and reassemble.
2. Check for an oil buildup, or tar-like substance on engine pulley.
   a. Clean the area thoroughly with plain regular gasoline.
3. If the drive disc is worn, or has flat spots on surface, replace it.
4. If chain case bearing is contacting cross brace bar, move cross bar up slightly.
   a. Use caution. If you move bar too far, it may allow metal part of drive disc to strike and ruin the engine drive wheel.

Adjusting Chain Drive Cam, the Drive Wheel and Bearing.

Chain adjustment is required if, mower will not drive, will not move in reverse, and seems to lack sufficient power to operate.

1. Maintain a distance of 3/8" between underside of drive cam support and bearing on the chain drive case, when the control is in either reverse or forward position. See figure 205.
2. To adjust, stand unit on aft end.
3. Press foot pedal down all the way and secure to foot rest.
   a. Use a large C-clamp which is available from rental agency.
4. Check distance and adjust right end of support to 3/8".
   a. Loosen the sideplate and move proper direction to adjust.
   b. Tighten the screws.
5. Remove the C-clamp and press rear of foot pedal down and hook it to the foot rest with a piece of wood.
6. Adjust left end of cam drive support to 3/8".
7. Remove piece of wood and test.
8. A clearance of .030" must be maintained between drive wheel, and engine pulley when chain is in neutral position.
   a. To adjust, set chain drive in neutral.
   b. Loosen mounting screws at center of cam support as needed.
   c. Use a feeler gauge to get proper setting between gear & pulley. See figure 206, on following page.
   d. Move cam support in proper direction to adjust.
   e. Tighten securely in place.
9. Use caution when making adjustment and be certain of accuracy.

Figure 205

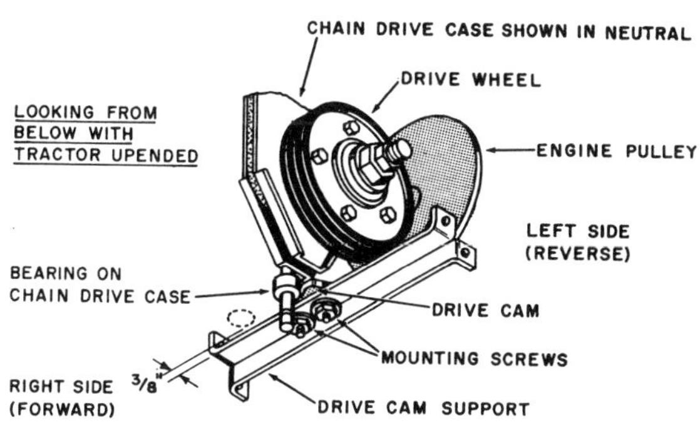

Figure 206

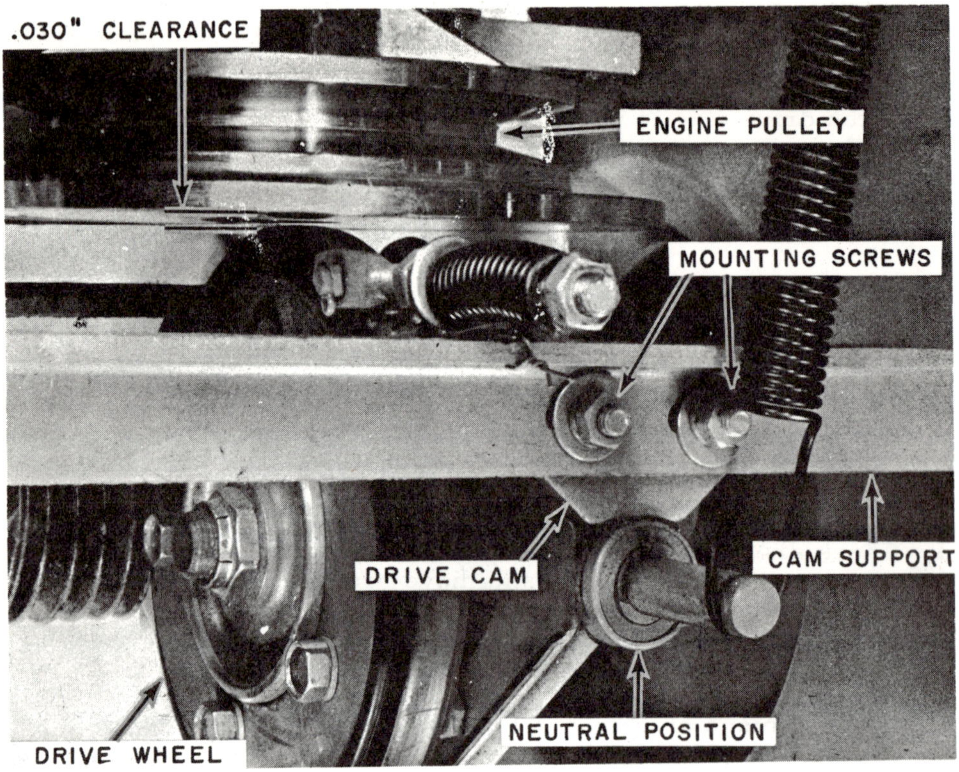

Brake Adjustment.

1. Although brake is factory adjusted, unit does require service.
2. If brake allows mower to coast back, the slack take-up spring isn't sufficiently compressed.
3. Stand mower on aft end.
4. The adjustment spring is located on parking brake rod(fig. 207a).
5. Turn nut at forward end of brake rod, clockwise, to tighten spring.
   a. Make adjustments in 1/4 turn increments.

Free Wheeling.

1. To move mower when engine isn't running, disengage wheel hub, as in figure 207.

Figure 207

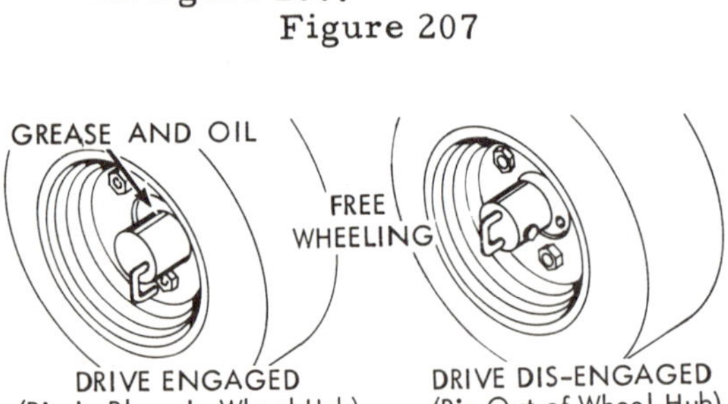

Figure 207a

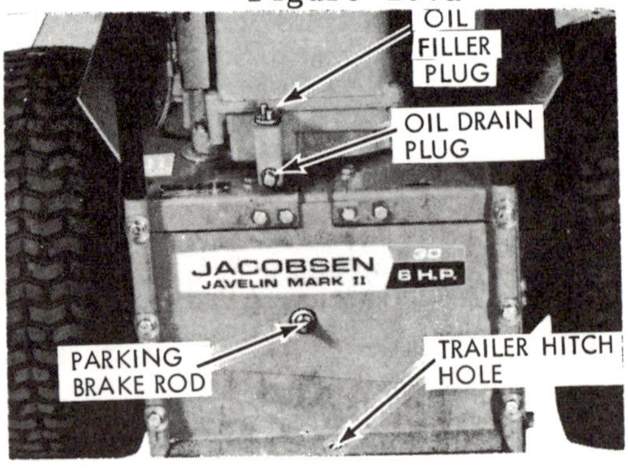

Chapter Seven

Ariens Manufacturing Company

The Ariens Company "Fairway" is a rider-type mower which incorporates either a 5 or 6 hp engine, depending on model. It also has an optional electric starting feature, in addition to the standard recoil starter.

Operating Instructions.

1. Before starting engine, read engine manual and check all pre-start procedures.
2. Familiarize yourself with the unit and all controls.
3. Place shift lever in neutral position.
4. Place clutch in OUT position.
5. Set throttle control on CHOKE position. When engine starts, move control to SLOW position. If engine is warm, try starting without choke.
6. On recoil-equipped models, start engine by pulling sharply on the starter handle.
7. To stop engine, set throttle control in STOP position.
8. On electric starter models, start the engine by turning key switch to start position.
    a. When engine starts, turn switch to RUN position.
    b. To stop engine, turn key to OFF position.
    c. Do not pull recoil starter with battery connections off.

Figure 208

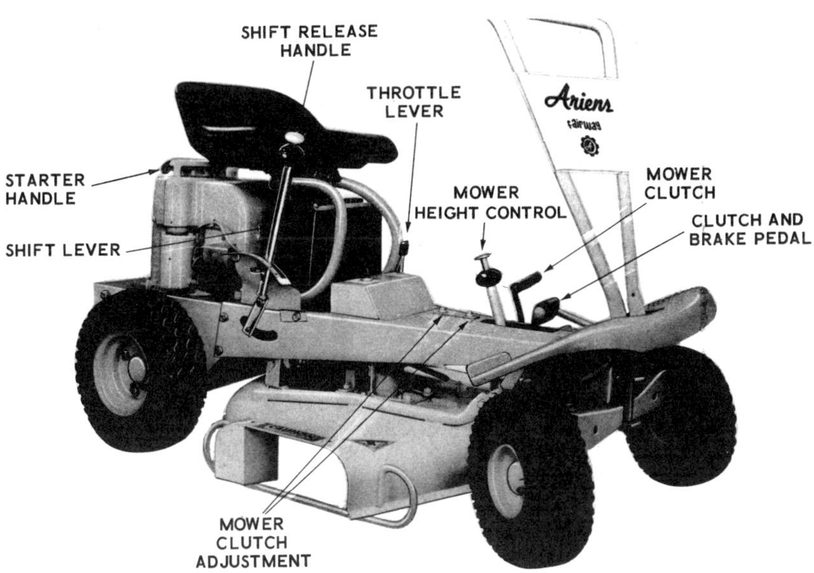

Operation.

The unit features a combination automotive-type clutch and brake pedal which disengages clutch part-way down and enables shifting to four speeds forward and one reverse. When pedal is depressed beyond shifting range, it becomes the brake.

1. To operate, depress clutch pedal, depress shift release handle, and move shift handle to desired speed, and release the clutch slowly.
2. When tractor is moving, shifting can be performed without the clutch.
3. Clutch must be engaged when shifting back to neutral or in reverse.
4. Always remove your foot from clutch pedal after shifting.
5. Forward speeds are 1.5, 2.5, 3 and 4 mph forward, and 2 mph in reverse.

Mower Clutch.

1. The clutch for engaging and dis-engaging the mower is a different control. See figure 208.
2. To start mower, move clutch pedal forward. To stop, move clutch pedal to the rear.
3. The mower contains a special feature, a blade brake which is attached to the mower pan and engages a brake drum on the mower sheave to stop the mower when the clutch is disengaged.
4. When adjustment is required, refer to figure 208, and notice bolts in slotted clutch bracket and frame.
5. If mower belt is loose and slips when clutch is engaged, loosen the two screws and slide clutch bracket toward rear of mower.
   a. Tighten the screws and test. Adjust further if needed.
6. If mower belt is too tight and clutch will not lock in engaged position, loosen the screws holding clutch bracket and move forward.
   a. Tighten screws and test. Adjust further if needed.
7. If belt is worn and requires replacement, place mower clutch pedal in the rear or OUT position.
8. Place mower height control in lowest cutting position.
9. Use a block to prop rear of mower deck away from tractor frame.
10. Rotate blade spindle by hand and roll belt off mower sheave until it is between sheave and belt finger. Belt finger will move away from the sheave so belt can be removed.
11. Remove belt finger(figure 209) by removing cap screw and nut from side of frame.
12. Place shift lever in fourth speed and remove cap screw opposite the belt finger.
13. Place shift lever in reverse and remove belt finger. Remove belt from sheave by rolling it over sheave, while rotating the blade.
14. To replace, position belt on engine sheave, then on mower sheave.
15. Replace engine belt finger.
    a. Finger must clear belt around by 1/8".
16. Readjust the clutch if necessary.

Drive Belt Adjustment.

1. Remove the mower belt per above instructions.
2. Use a screwdriver and small bar to pry idler pulley against drive disc, and slide belt around the idler pulley.

3. Belt should be loose. Place throttle control in STOP position.
4. Rotate engine pulley by pulling on recoil starter. Use slight pressure on belt and roll it away from the drive disc.
5. Place shift lever in NEUTRAL position.
6. Use a 5/8" open end wrench to rotate hex shaft and friction wheel, while pulling belt upward between drive disc and friction wheel.
7. Replace with a similiar belt. It's a good idea to use original belt from manufacturer if available.
8. Shift control should remain in NEUTRAL position as you slide the belt in from top of frame.
9. Use same wrench on hex shaft and rotate friction wheel to force belt between drive disc and the wheel.
10. Shift to REVERSE position and place belt around back groove of the engine pulley.
11. Shift to NEUTRAL position and rotate the belt around drive disc.
12. Use a screwdriver and small bar to move idler pulley toward drive disc, while moving the belt in position outside of the idler pulley.
13. Replace mower belt and belt finger per instructions on previous page.

Cutting Height Control.

1. Cutting height of mower is adjustable from 1 1/2" to 3 1/4" by changing height control arm.
2. To raise the mower blade, depress height control plunger on height control knob and push the arm forward until desired height is attained, then allow control plunger to spring up and lock the arm in position.
3. To lower the blade, depress height control plunger, pull back the control arm until it locks in position.
   a. It may be necessary to push height control arm forward slightly to relieve pressure on control plunger before plunger can be easily depressed.

Figure 209

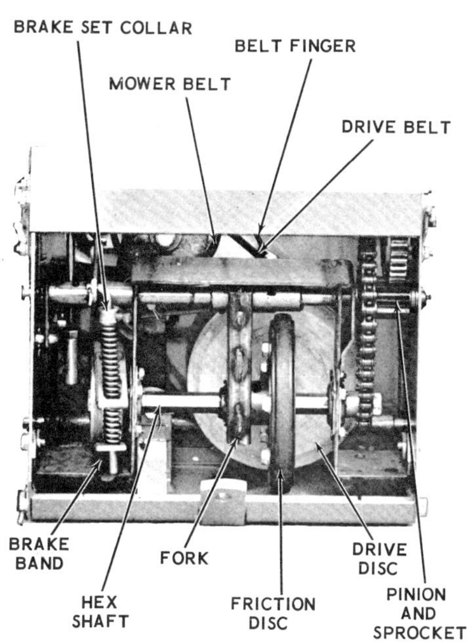

Figure 210

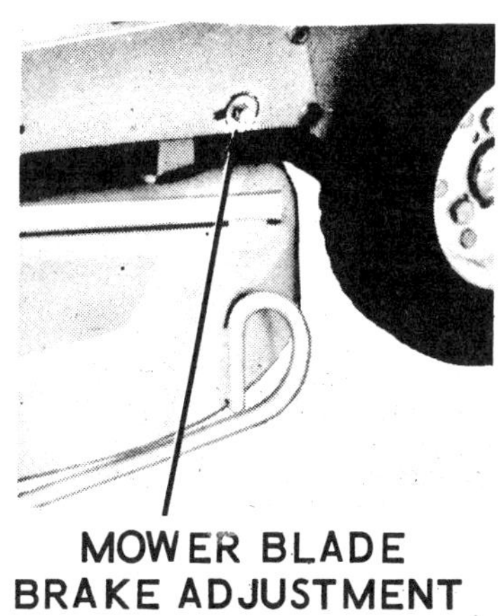

Mower Blade Brake Adjustment.

1. The blade brake release (figure 210) is located on left side of frame and is held by a 5/16" screw. The blade brake release must be properly positioned so the brake will stop mower blade when the clutch is released.
2. Set mower clutch pedal in rear, or OUT position.
3. Adjust mower to highest cutting position.
4. Loosen screw holding blade brake release and slide release back until brake release chain barely touches the top of mower pan, and tighten the screw.
5. Engage mower pedal to be sure blade brake releases as belt begins to tighten.

Brake Adjustment.

1. Set shift lever in NEUTRAL position.
2. Adjust collar (figure 209) against spring until brake band just clears brake drum. Turn it inward.
3. Some brake drag isn't harmful. Check for too much drag by turning both wheels, by hand, at the same time.

Friction Disc Replacement.

The friction disc will wear after prolonged usage and replacement will be required.

1. Set shift control in NEUTRAL position.
2. Remove both cap screws holding carrier bracket to frame. Remove the 5 screws holding friction disc to the hub.
3. Remove cap screw and lockwasher holding ball joint to sliding fork.
4. Pull ball joint away from fork.
5. Remove cotter pin and disconnect brake rod from the clutch brake bellcrank.
6. Pull the brake rod from brake band and remove both springs, the set collar and brake band washer.
7. Swing lower end of carrier bracket forward away from frame, and tie, or prop it up.
8. Hold hex shaft with a wrench and remove the 1/2" locknut holding the brake drum.
9. Pull brake drum away from shaft and remove shaft key.
10. Disconnect the link on roller chain and remove the chain.
11. Remove four locknuts holding bearing flanges on sprocket end of hex shaft. Pull carrier bracket forward and remove hex shaft from sprocket end. Also remove from bearing on brake drum end and pull out of the friction disc hub. The bearing on sprocket end will pull out of carrier.
    a. Pull out just enough to allow removal of friction disc and hub.
12. To install, place new friction disc on hub but don't bolt in place, and slide hub into place in the fork.
13. Slide hex shaft through the hub and into place, in the bearing, on the brake drum end.
14. Replace bearing flanges on the sprocket end of the shaft and secure with the four locknuts.
15. Replace the roller chain and link.
    a. Follow steps above for replacing or repairing chain.
16. Place key on shaft and replace brake drum. Tighten 1/2" locknut.

17. Tip carrier back into position and secure on each side with cap screw through frame and into the carrier.
18. Reconnect ball joint to sliding fork.
19. Replace brake washer, springs, set collar and brake rod. Replace the cotter pin.
20. Position friction disc on the hub and replace the 5 screws.

Shifting.

1. If shifting becomes harder after prolonged use, or immediately after replacing friction disc, oil shift lever and tube, and clean and re-oil the hex shaft which holds firction disc, and grease the hub.
2. Do not oil or grease friction or drive discs.

Lubrication.

1. Oil should be applied to shift lever pivot, mower clutch lever, the steering column, and all pivot points.
2. Oil the chain frequently.
3. Grease the bell crank, sliding fork, pinion and sprocket, sliding hub, and L. H. axle at least every 25 hours. Oftener under extreme conditions. See figure 211.
4. Lubricate and change oil per engine instruction manual.
5. Front wheels should be greased at least once a season.

Battery Care(Electric Starting Models).

1. Battery connections should be tight at all times.
2. When corrosion results, remove cables and clean with a wire brush or a solution of one part baking soda, to four parts water. Clean both cable and terminals, but don't allow solution to enter battery cells.
3. Tighten securely when replacing. Coating terminals with petroleum jelly will reduce corrosion problem.
4. Check water level of each cell. Water should be level with top of circulating split ledge inside filler tube. Refill with distilled water if available.
5. Keep liquid at proper level during winter or freezing weather. Also keep battery fully charged to prevent damage to the battery.
6. To re-activate, leave battery indoors in temperature of 60 degrees to 110 degrees F. Add electrolyte to battery before installing on mower.
7. Follow instructions when handling the acid and avoid spillage.
8. Fill each cell until acid level reaches triangle in vent well. After all cells are filled, recheck and add electrolyte if needed.
   a. Use caution and don't overfill.
9. Charge battery at 10 AMPS until electrolyte temperature is 80 de-

grees or higher and specific gravity to 1.250 or higher. Lower charging rate may be used to obtain the temperature and specific gravity, but it will take longer.

10. After battery is charged, cleanse with water to remove acid overflow.

11. Best performance is attained when battery is kept on or near full charge.

12. When handling acid, exercise extreme caution. Avoid getting the solution on your body or into your eyes. Use the safety glasses if they are available.

a. If acid gets on your person, or internally, follow instructions printed on acid container for an antidote.

Whether you are working on a customers machine, or your own, always follow proper safety rules. If you're in doubt, read final chapter on safety rules and regulations.

Figure 212

Cutter Blade and housing components.

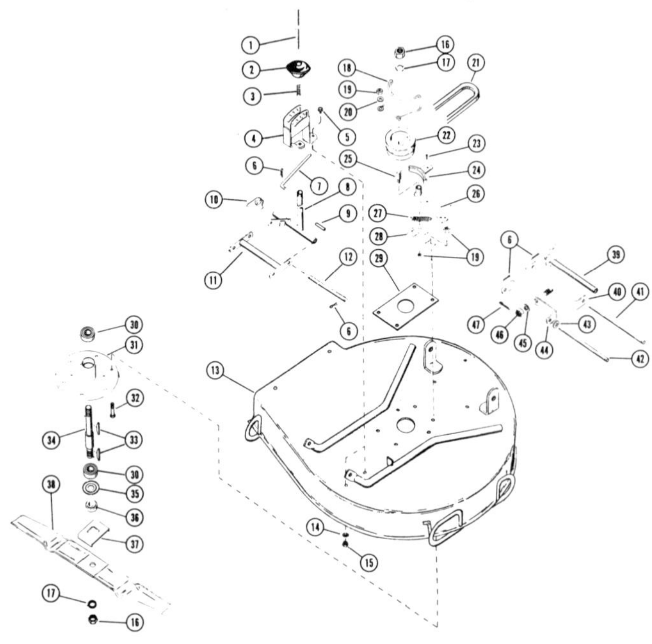

REF. NO.	PART NO.	DESCRIPTION	NO. REQ'D.	REF. NO.	PART NO.	DESCRIPTION	NO. REQ'D.
1	12056	Height Control Plunger	1	26	12111	Blade Brake Arm	1
2	10248	Shift Ball	1	27	83010	Spring	1
3	83015	Spring	1	28	12062	Blade Brake Support	1
4	12170	Mower Height Quadrant	1	29	3117	Reinforcing Plate	1
5	59022	Screw 5/16"-18 x 3/4"	2	30	54073	Bearing	2
6	67004	Cotter Pin 1/8" x 1"	5	31	3172	Spindle Housing	1
7	12117	Height Link Rod	1	32	59004	Cap Screw 3/8"-16 x 1" H.H.	6
8	12112	Height Control Arm	1	33	66003	Key, #9 Woodruff	2
9	58007	Roll Pin 1/4" x 1-1/4"	1	34	3127	Mower Spindle	1
10	83027	Spring	1	35	3419	Bearing Slinger	1
11	12113	Floating Link	1	36	3169	Retainer Hub	1
12	12125	Adapter Pin	1	37	3170	Blade Tray	1
13	12165	Mower Pan	1	38	3165	Mower Blade	1
14	63003	Lockwasher	2	39	12116	Rear Raising Bracket	1
15	65015	Locknut 5/16"-18 Plated	2	40	12115	Clevis	1
16	65025	Hex Jam Nut 3/4"-10 Plated	2	41	12059	Rear Bracket Pin	1
17	63008	Lockwasher 3/4"	2	42	12067	Rear Hanger Rod	1
18	12061	Mower Belt Finger	1	43	64003	Wrought Washer 1/2"	1
19	65039	Locknut 3/8"	7	44	12114	Clevis	1
20	69002	Washer	2	45	64039	Washer 1.005 O.D. x .630 I.D. x .062	1
21	72029	V-Belt	1	46	65049	Nut 5/8"-18 Slotted Jam	1
22	12025	Mower Sheave	1	47	67014	Cotter Pin 1/8" x 1-1/4"	1
23	68016	Rivet 1/4" x 1/2" Rd. Hd.	1			(NOT ILLUSTRATED)	
24	12063	Blade Brake	1		78160	Safety Label	1
25	3016	Shoulder Bolt	1		78028	Discharge Caution Label	1

Figure 213
Components and nomenclature for the "Fairway" Rider Mower

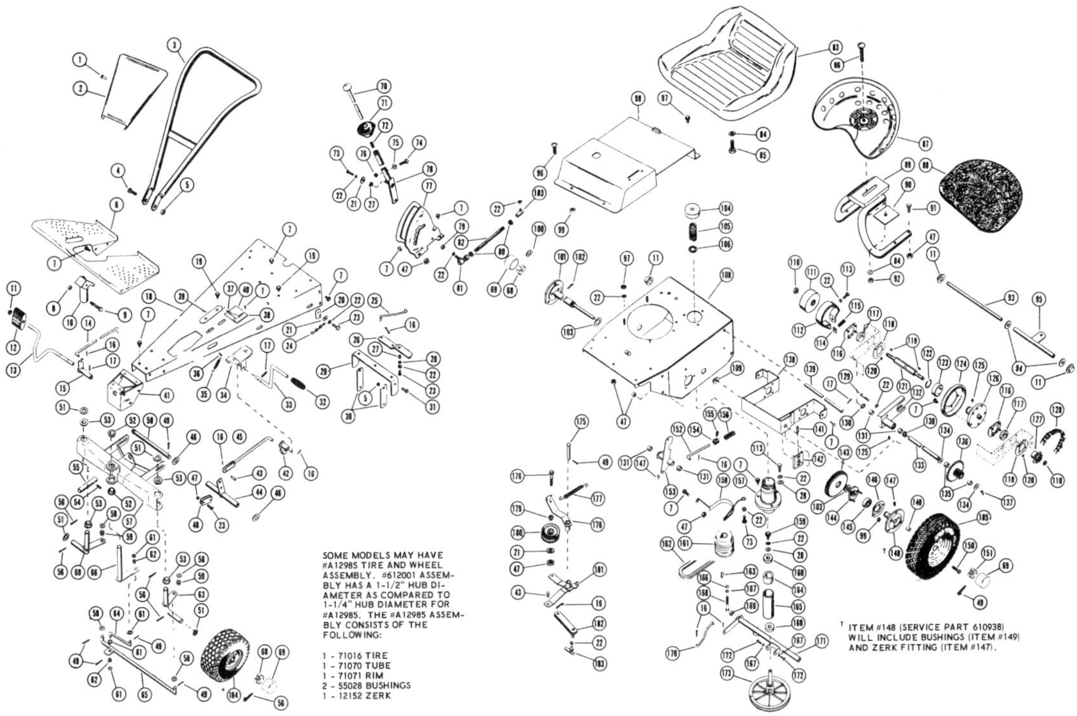

REF. NO.	PART NO.	DESCRIPTION	NO. REQ'D. 912970	912971	912972
1	68031	Pop Rivet	6	6	6
2	12158	Handle Bar Panel	1	1	1
3	12162	Handle Bar	1	1	1
4	59035	Cap Screw 3/8"-16 x 2-1/2" H.H.	1	1	1
5	65039	Locknut 3/8"-16 Hex	3	3	3
6	12187	Footrest	1	1	1
7	70009	Screw, Flange Whizlock 5/16"-18 x 12"	25	25	24
8	65040	Locknut 1/4"-20	1	1	1
9	59079	Cap Screw 1/4"-20 x 1-1/2" H.H.	1	1	1
10	12106	Steering Block	1	1	1
11	12086	Cap	4	4	4
12	12085	Foot Grip	1	1	1
13	12146	Foot Pedal	1	1	1
14	12040	Front Clutch Rod	1	1	1
15	12041	Internal Clutch Lever	1	1	1
16	67001	Cotter Pin 3/32" x 3/4"	9	9	9
17	58016	Roll Pin 3/16" x 1"	5	5	5
18	12082	Frame Channel	1	1	1
19	70010	Screw, Flange Whizlock 3/8"-16 x 1/2"	5	5	5
20	12109	Brake Release Bracket	1	1	1
21	64002	Washer 5/16" Std.	3	3	3
22	63003	Lockwasher 5/16" SAE Std.	12	12	12
23	59022	Cap Screw 5/16"-18 x 3/4" H.H.	4	4	4
24	12127	Brake Release Chain	1	1	1
25	12173	Rear Clutch Rod	1	1	1
26	12049	Clutch Transfer Lever	1	1	1
27	6071	Spacer	2	2	2
28	64007	Washer 1/4" Std.	3	3	3
29	12050	Hanger Bracket	1	1	1
30	12124	Hanger Link	2	2	2
31	3316	Shoulder Bolt	2	2	2
32	75004	Grip	1	1	1
33	12126	Mower Clutch Pedal	1	1	1
34	12046	Mower Clutch Bracket	1	1	1
35	83025	Spring	1	1	1
36	58025	Groove Pin	1	1	1
37	75019	Knob	1	1	1
38	69058	Throttle Control	1	1	1
39	12090	Guide Plate	1	1	1
40	12177	Brand Plate	1	1	
41	12107	Front Frame	1	1	1
42	12047	Mower Overcenter Lever	1	1	1
43	12151	Clevis Pin	2	2	2
44	12108	Mower Clutch Lever	1	1	1
45	12045	Mower Clutch Link	1	1	1
46	64003	Washer 1/2" SAE	2	2	2
47	65042	Locknut 5/16"-18 Hex	14	14	14
48	12044	Front Link	2	2	2
49	67004	Cotter Pin 1/8" x 1"	7	7	7
50	12067	Hanger Rod	1	1	1
51	64048	Washer 1.375 O.D. x .755 I.D. x .062	4	4	4
52	55038	Flange Bushing	2	2	2
53	12128	Bushing	4	4	4
54	58007	Roll Pin 1/4" x 1-1/4"	1	1	1
55	12021	Axle Pivot Pin	1	1	1
56	67014	Cotter Pin 1/8" x 1-1/4"	6	6	6
57	12101	Front Axle	1	1	1
58	64038	Washer .775 O.D. x .505 I.D. x 24 Ga.	4	4	4
59	55037	Flange Bushing	2	2	2
60	12103	R. H. Spindle	1	1	1
61	64052	Washer 13/16" O.D. x .380/.385 I.D. x 24 Ga.	4	4	4
62	55036	Flange Bushing	2	2	2
63	12102	L. H. Spindle	1	1	1
64	12145	Steering Link	1	1	1
65	12105	Tie Rod	1	1	1
66	12104	Steering Post	1	1	1
68	3376	Spindle Cup	3	3	3
69	3377	Hub Cap	4	4	4

REF. NO.	PART NO.	DESCRIPTION	NO. REQ'D. 912970	912971	912972
70	3247	Shift Release Handle	1	1	1
71	10248	Shift Ball	1	1	1
72	83015	Spring	1	1	1
73	59003	Cap Screw 5/16"-18 x 1" H.H.	2	2	2
74	59072	Cap Screw 1/4"-20 x 1-1/4" H.H.	1	1	1
75	64037	Washer 1-1/4" O.D. x 9/32" I.D. x 1/16"	1	1	1
76	65070	Locknut 1/4"-20	1	1	1
77	12075	Shift Control Segment	1	1	1
78	3089	Shift Lever	1	1	1
79	65015	Nut 5/16"-18 Hex	3	3	3
80	65031	Nut 5/16"-24 Hex	2	2	2
81	3060	Ball Joint	1	1	1
82	12172	Shift Link	1	1	1
83	3375	Seat Bantam			1
84	63006	Lockwasher 1/2" SAE Std.	1	1	1
85	59042	Cap Screw 1/2"-13 x 1-1/4"			1
86	62014	Carriage Bolt 1/2"-13 x 1-1/4"	1	1	
87	12083	Seat	1	1	
88	12175	Seat Cover	1	1	
89	12189	Seat Bracket	1	1	1
90	12079	Seat Support Washer	1	1	1
91	59045	Cap Screw 5/16"-18 x 1-1/2" H.H.	4	4	4
92	65021	Nut 1/2"-13 Hex	1	1	
93	3159	Hitch Shaft	1	1	1
94	3254	Wave Washer	2	2	2
95	3252	Hitch	1	1	1
96	61032	Machine Screw 1/4"-20 x 5/8"			2
97	70017	Screw, Flange Whizlock 5/16"-18 x 5/8"			1
98	12153	Dash Panel			1
99	65056	Nut, Flange Whizlock 1/4"-20	6	6	8
100	10366	Pushnut	1	1	1
101	12171	R. H. Axle	1	1	1
102	58003	Roll Pin 3/16" x 1-1/4"	2	2	2
103	64036	Washer 1.380 O.D. x .880 I.D. x .062	2	2	2
104	12013	Cap	1	1	1
105	83055	Spring	1	1	1
106	12014	Spring Retainer	1	1	1
108	12174	Rear Frame	1	1	1
109	12129	Bushing	2	2	2
110	65054	Locknut 1/2"-20 Hex Jam	2	2	2
111	12026	Brake Drum	1	1	1
112	12033	Brake Band	1	1	1
113	59039	Cap Screw 5/16"-18 x 1/2" H.H.	2	2	2
114	12074	Brake Band Washer	1	1	1
115	83028	Spring	1	1	1
116	3085	Bearing Flange	2	2	2
117	54080	Bearing	2	2	2
118	3031	Bearing Flange	2	2	2
119	66015	Key, Woodruff #5	2	2	2
120	65051	Nut 10-24	8	8	8
121	12029	Hex Shaft	1	1	1
122	57043	Snap Ring	1	1	1
123	54083	Bearing	1	1	1
124	3003	Friction Wheel	1	1	1
125	12152	Zerk Fitting	2	2	2
126	12183	Friction Disc Hub	1	1	1
127	10276	Sprocket	1	1	1
128	12078	Chain	1	1	1
129	59036	Cap Screw 5/16"-24 x 1/2" H.H.	1	1	1
130	56034	Seal	2	2	2
131	55047	Bushing	4	4	4
132	12095	Sliding Fork	1	1	1
133	12181	Carrier Shaft	1	1	1
134	64039	Washer 1.005 O.D. x .630 I.D. x .062	2	2	2
135	54038	Needle Bearing	2	2	2
136	12141	Pinion and Sprocket	1	1	1
137	13136	Zerk Fitting	1	1	1
138	12027	Carrier	1	1	1
139	12039	Disc Fork Rod	1	1	1

REF. NO.	PART NO.	DESCRIPTION	NO. REQ'D. 912970	912971	912972
141	58030	Roll Pin	1	1	1
142	12180	Locating Plate	1	1	1
143	612999	Differential Assembly	1	1	1
144	3017	Bearing Flange	2	2	2
145	54093	Ball Bearing	2	2	2
146	10142	Bearing Flange	2	2	2
147	10354	Zerk Fitting	2	2	2
148	010938	L. H. Axle	1	1	1
149	55029	Bushing	2	2	2
150	70005	Wheel Bolt	6	6	6
151	12190	Spindle Cup	1	1	1
152	12032	Brake Rod	1	1	1
153	12182	Clutch Brake Bellcrank	1	1	1
154	3074	Set Collar	1	1	1
155	60018	Setscrew 1/4"-20 x 1/2" Sq. Hd.	1	1	1
156	83013	Brake Spring	1	1	1
157	12006	Drive Spring Housing	1	1	1
158	12147	Rear Belt Finger	1	1	1
159	70014	Cap Screw, Special	1	1	1
160	54039	Ball Bearing	2	2	2
161	12024	Engine Sleeve	1	1	1
162	72048	V-Belt	1	1	1
163	66009	Key, Woodruff #6	1	1	1
164	12100	Bearing Spacer	1	1	1
165	12010	Drive Spindle Housing	1	1	1
166	64028	Washer 1-1/8" O.D. x 7/16" I.D. x 3/16"	1	1	1
167	63004	Lockwasher 3/8" SAE Std.	3	3	3
168	59017	Cap Screw 3/8"-24 x 1-1/4" H.H.	1	1	1
169	10132	Fork Spacer	1	1	1
170	12030	Clutch Link	1	1	1
171	12144	Disc Fork	1	1	1
172	10522	Screw Pin	2	2	2
173	612003	Drive Spindle Assembly	1	1	1
175	12150	Pin	1	1	1
176	59069	Cap Screw 5/16"-18 x 1-1/4" H.H.	1	1	1
177	83026	Idler Spring	1	1	1
178	12142	Idler Arm	1	1	1
179	12131	Bearing Spacer	1	1	1
180	12132	Idler	1	1	1
181	12094	Shift Transfer Lever	1	1	1
182	12096	Shift Link	1	1	1
183	3086	Ball Joint	2	2	2
184	612005	Tire and Wheel Assembly	2	2	2
		Consisting of:			
		1 - 71085 Tube			
		1 - 71080 Tire			
		1 - 71083 Rim or 71084 Rim (71084 RIM HAS VALVE STEM HOLE ON ZERK FITTING SIDE. 71083 OPPOSITE.)			
		*2 - 55058 Bushings			
		1 - 12152 Zerk Fitting			
		*NOTE: 54089 bearings can be used in place of 55058 bearings			
185	612002	Tire and Wheel Assembly	2	2	2
		Consisting of:			
		1 - 71052 Rim			
		1 - 71081 Tire			
		1 - 71050 Tube			
		(NOT ILLUSTRATED)			
	78002	Height Quadrant Label	1	1	1
	78003	Mower In-Out Label	1	1	1
	78017	Pedal Label	1	1	1
	78018	Mower Shift Label	1	1	1
	78019	Engine Label	1	1	1
	78034	Engine Decal, 5 H.P.	1		
	78035	Engine Decal, 6 H.P.		1	1
	78083	Caution Label	1	1	1
	78154	Brake Decal	1	1	1
	78155	Brake Decal	1	1	1
	78150	Brake Decal			1

Figure 214
Electrical System illustrating complete wiring diagram.

ELECTRICAL SYSTEM

MODEL 12972

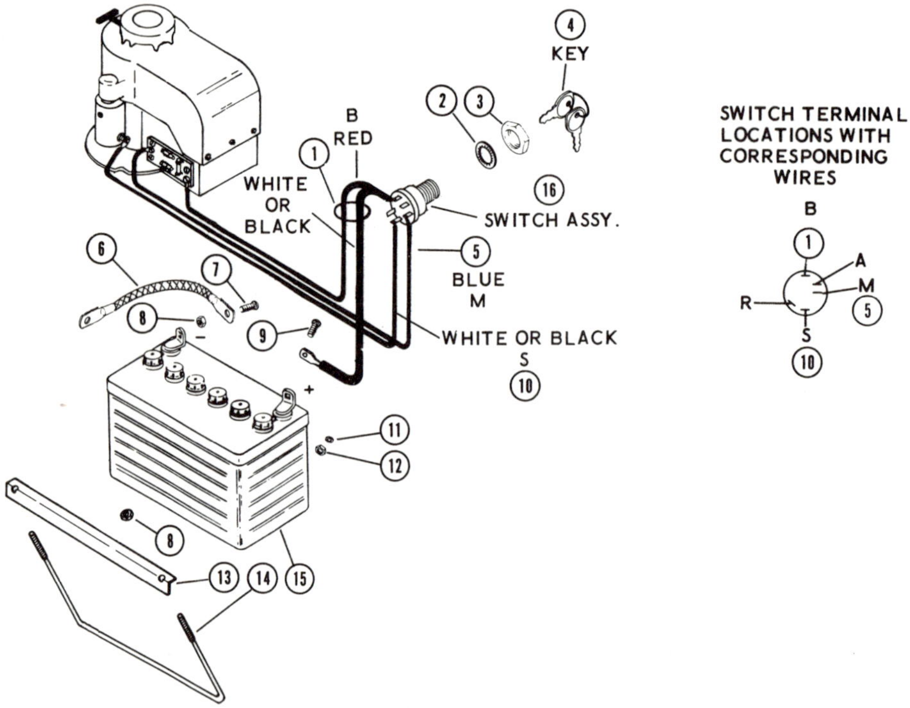

REF. NO.	PART NO.	DESCRIPTION	NO. REQ'D.
1	12168	Positive Battery Cable	1
2	63019	Lockwasher	1
3	65075	Nut	1
4	13157	Key (only)	2
5	12169	Shorting Cable	1
6	3323	Ground Cable	1
7	59080	Cap Screw 1/4"-20 x 5/8" H.H.	2
8	65056	Nut, Flange Whizlock 1/4"-20	4
9	62011	Carriage Bolt 5/16"-18 x 3/4"	1
10	12166	Starter Cable	1
11	63003	Lockwasher 5/16" Std.	1
12	65015	Nut 5/16"-18 Hex	1
13	13044	Battery Hold Down	1
14	12155	Battery Clamp	1
15	13081	Battery	1
16	613943	Ignition Switch Assembly	1
		Note: Parts 3 and 4 are furnished when switch is ordered.	

Chapter Eight

Sensation Manufacturing Company

The Sensation Company, Ralston, Nebraska, manufacture a complete line of power rotary lawn mowers. They use mostly the Briggs & Stratton engine with several innovations of their own.

The MOW-BLO 21G, illustrated below, incorporates a feature which is designed to vacuum clean as it mows.

Figure 215

The Sensation line of mowers feature several innovations that place the company with other leaders of the industry.

1. The Magic Cone Crankshaft Protector is a cone-shaped protector, that acts as an additional bearing at the bottom of crankshaft and almost eliminates bending. See figure 216.
2. Figure 217 illustrates two blade options. The air lift aids in vacuum feature of Mow-Blo units.
3. The patented Gyro(r) blade is an option worth consideration.
4. The gyroscopic rotary baffle balance wheel can be used with swing-back safety blades on outside edges. With the two blade types over 28 different combinations can be made.
    a. The system helps improve cutting uniformity.
    b. Helps eliminate crankshaft damage.
    c. Makes engine starting easier.
    d. Swing-back blades reduce blade damage and reduce service.
    e. The lightweight blades do more cutting at less engine speed.

Another innovation used on several Sensation mowers is a modified valve system. It is purported to increase efficiency and double valve life.

With valve in closed position, the Belleville spring is loaded at point(1) and has a reaction at point(2). As the valve opens, the additional spring-load on valve causes the washer-type spring to flatten, which transfers the reaction load from point(2) to point(3), forcing the balls to roll down the inclined races. Reaction load of the balls on the inclined races induces rotation to the retainer, causing a torque to be applied to the valve. As the valve closes, Belleville spring is released from the balls, allowing them to return to their original position.

Figure 216

Figure 218

Figure 219

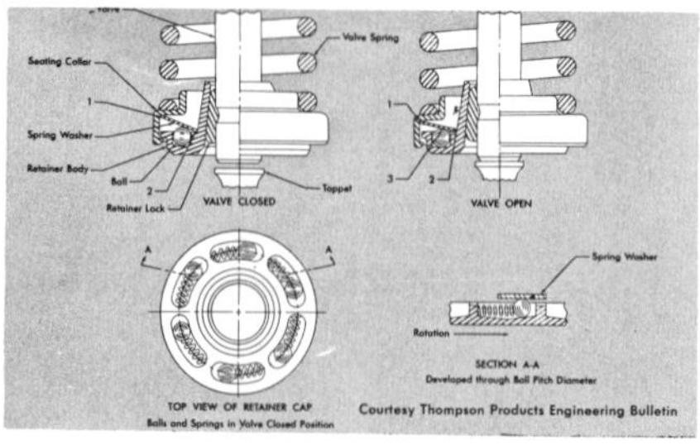

Figure 217

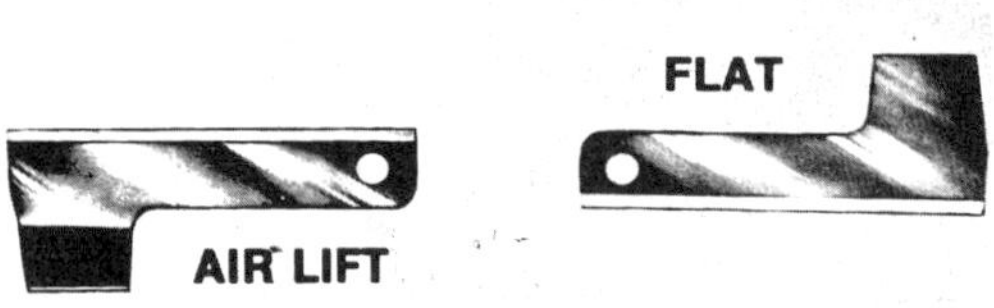

Chapter Nine

AMF Western Tool Division

AMF units feature many new innovations designed to make lawn mowing simpler. They utilize mostly Briggs & Stratton engines with many refinements including, top oil fill, visual gas gauge and electric starter.

Height adjustment controls are simple and some models have increment dials adjacent to adjusting lever. See figure 220.

The reel-type models also feature a simple reel-bed knife adjustment.

Figure 220

Figure 221

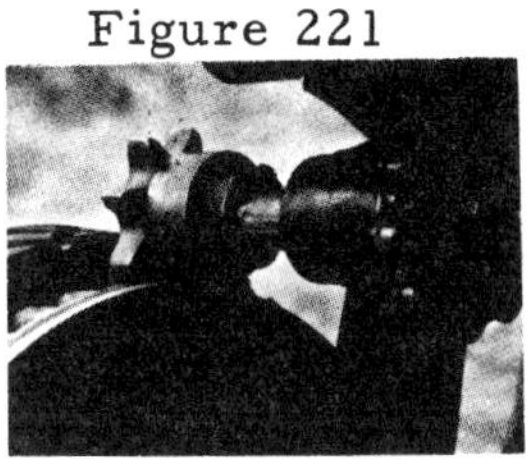

AMF's entry in the new cutter blade concept race is the "FLEXOR". It is hinged and bends when it strikes an object, almost eliminating the shock and damage to engine crankshaft. In fact, AMF guarantees every mower containing the "Flexor" blade as original equipment, and will replace any damaged unit components when returned to AMF factory intact, with-out charge.

Self-propelled units feature a heavy front-wheel gear drive, which aids in tilt-back turning. Service is also simple and open. Figure 221

Underside of aluminum cast deck illustrating the "Flexor" cutter blade.

Figure 222

Specification sheet on the AMF line of lawn care equipment.

MODEL NO.	DESCRIPTION	H.P.	START	HEIGHT ADJUSTMENT	BLADE	HANDLE	GRASS CATCHER	GAS GAUGE	TIRE SIZE FRONT	TIRE SIZE REAR
1218	18″ Trimmer Push	3.0	Stand-Up	Quick Wheel	Flexor	Amfold	Optional	No	7″	7″
1250	20″ Standard Push	3.0	Pull-Up	Bolt	Flexor	Amfold	Optional	No	7″	7″
1252	20″ Deluxe Push	3.5	Pull-Up	Single Point	Flexor	Amfold	Standard	Yes	7″	7″
1253	22″ Deluxe Push	3.5	Pull-Up	Single Point	Flexor	Amfold	Standard	Yes	8″	8″
1254	22″ Deluxe Self Propelled	3.5	Pull-Up	Single Point	Flexor	Amfold	Standard	Yes	8″	8″
1255	22″ Deluxe Self Propelled	3.5	Electric Start	Single Point	Flexor	Amfold	Standard	Yes	8″	8″
1256	22″ Deluxe Push	3.5	Electric Start	Single Point	Flexor	Amfold	Standard	Yes	8″	8″

AMF CAST DECK ROTARIES

MODEL NO.	DESCRIPTION	H.P.	START	HEIGHT ADJUSTMENT	BLADE	HANDLE	GRASS CATCHER	GAS GAUGE	TIRE SIZE FRONT	TIRE SIZE REAR	WASH-OUT PORT SAFETY SHIELD
1257	21″ Deluxe Push	3.5	Pull-Up	Quick Wheel	Flexor	Amfold	Standard	Yes	8″	8″	Yes
1258	21″ Deluxe Self Propelled	3.5	Pull-Up	Quick Wheel	Flexor	Amfold	Standard	Yes	10″	8″	Yes
1259	21″ Deluxe Self Propelled	3.5	Electric	Quick Wheel	Flexor	Amfold	Standard	Yes	10″	8″	Yes

AMF SODBUSTER TILLERS

MODEL NO.	DESCRIPTION	H.P.	START	REVERSE	TRANS-MISSION	DRIVE	TINES	TIRE SIZE
1344	22″ Deluxe Tiller	3.5 Vertical	Recoil	Yes	Ultragear	Belt	14″ Bolo	8″
1348	26″ Deluxe Tiller	4.0 Horizontal	Recoil	Yes	Ultragear	Belt	14″ Bolo	10″
1349	28″ Deluxe Tiller	5.0 Horizontal	Recoil	Yes	Ultragear	Belt	14″ Bolo	10″

AMF RIDING MOWERS

MODEL NO.	DESCRIPTION	H.P.	START	HEIGHT ADJUST-MENT	BLADE	BLADE CLUTCH	TRANS-MISSION	PARK-ING BRAKE	GRASS CATCHER	TIRE SIZE FRONT	TIRE SIZE REAR
1263	26″ Mow Trac	5.0	Recoil	Single Lever	Single	Yes	4-Speed Orbit Disc	Yes	No	10.5 x 3.5	15 x 6.00
1266	30″ Mow Trac	8.0	Recoil	Single Lever	Single	Yes	4-Speed Orbit Disc	Yes	Optional	11 x 4.00	16 x 6.50
1268	30″ Mow Trac	8.0	Electric	Single Lever	Single	Yes	4-Speed Orbit Disc	Yes	Optional	11 x 4.00	16 x 6.50
1272	26″ Tractor Look	8.0	Recoil	Single Lever	Single	Yes	3-Speed Gear	Yes	No	11 x 4.00	15 x 6.00
1278	26″ Tractor Look	8.0	Electric	Single Lever	Single	Yes	3-Speed Gear	Yes	No	11 x 4.00	15 x 6.00
1282	36″ AMF Trac	8.0	Recoil	Single Lever	Twin Blades	Yes	3-Speed Transaxle	Yes	No	15 x 6.00	18 x 8.50
1281	36″ AMF Trac	8.0	Electric	Single Lever	Twin Blades	Yes	3-Speed Transaxle	Yes	No	15 x 6.00	18 x 8.50

Chapter Ten

Electric Mowers

Pincor Model 20EB Electric Rotary Mower.

The electric mower is gaining in popularity, especially with homeowners in cities and close-quarter subdivisions with small lawns.

The Electric Motor.

The average do-it-yourselfer should never attempt motor repair unless he is well versed in the area.

1. If mower is plugged into electrical outlet and motor doesn't start, check the cord for breaks. Be sure cord is plugged in.
    a. If both of above check out, check connections of cord to motor.
    b. A motor will generally start malfunctioning gradually, and not just burn out on the spot. If a gradual loss of power has been taking place, take unit to electric repair shop for inspection.
2. Check your extension cord. Electricity dissapates as length of the cord increases. Longer cord requires heavier gauge wire.

50 to 100 feet cord	use 16 gauge wire
100 - 200 feet cord	use 14 gauge wire
200 - 300 feet cord	use 12 gauge wire

3. Cord must be 3-wire cable type with grounded plug outlet.
4. Always plan mowing route to avoid the cord.
5. Always wear shoes.

Height Adjustment.

1. To change cutting height, pull spring lever against wheel.
2. Rotate wheel and spring lever together to desired slot. Fig. 223.
3. To change height at rear wheels, lift mower off the ground.
4. Always unplug the unit when making adjustment.

Blade Service.

1. Refer to previous instructions on sharpening or replacing blade.

Lubrication.

1. Oil mower at point indicated in figure 224.
2. Other mower lubrication points are same as other Pincor units.

Figure 223

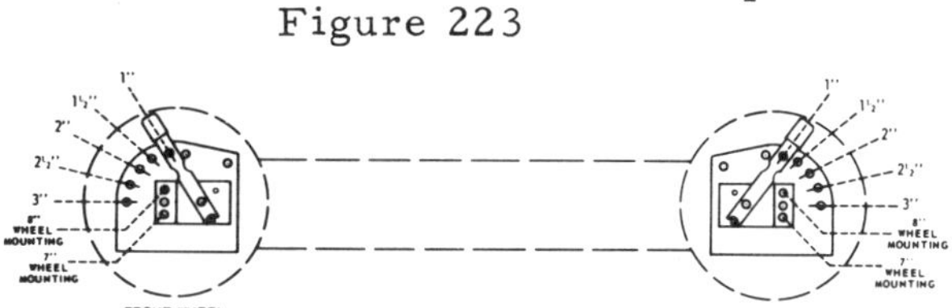

Figure 224

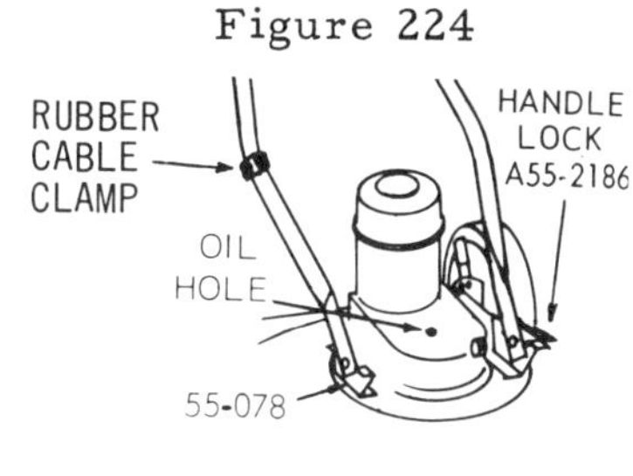

Figure 225
Complete component illustration and parts list for Model 20-EB mower.

COMPLETE PARTS LIST – MODEL 20-EB

Part No.	Name of Part
1016	Hex Nut
1076	Washer
1103	Screw
1295	Lockwasher
1341	Hex Screw 10-32 x 7/8
1590	Lockwasher 5/16 I.D.
1715	Lockwasher
1962	RD. HD. Screw 10-32 x 1-1/4
2493	Hex Nut 5/16-18
2681	Hex Screw 5/16-18 x 1" LG.
2719	Lockwasher 3/8 I.D.
3852	Lockwasher #10
10624	Hex Screw 1/4-20 x 5/8 LG.
10839	Screw
12217	Brush & Spring Ass'y.
13595	Screw
A51-036	Washer
55-052	Hex Locknut 1/4-20
55-078	Handle Brackets
55-109	Shoulder Bolts
55-159	Washer
55-404	Hex Bolt 3/8-24 x 1-1/2 LG.
55-406	Washer
55-411	Square Nut 3/8-16
55-551	Lockwasher 3/8 I.D.
55-713	Switch
55-841	Roll Pin
55-846	Adapter
A55-955	Spring Clip
55-998	Rubber Band
55-1011	Axles
55-1194	Washer
55-1272	Motor Mounting Base
55-1293	Hex Nut
55-1310	Cover
55-1580	Plug
55-1724	Wheel 8 x 1-3/4 Diamond Thread
55-1730	Chrome Hub Caps
55-1773	Blade 18"
55-1899	Hex Nut 10-32 Keps
55-1903	Shoulder Bolt
55-1904	Wheel Bracket
55-1907	Spring Lever Ass'y.
55-1908	Index Plate
55-1924	Chassis Assembly
55-1925	Subplate Ass'y.
55-1928	Cover & Hinge Ass'y.
55-1933	Extension
55-1951	Index Plate
55-1953	Wheel Bracket
55-1962	Sheet Metal Screw
55-1963	Screw Binder 10-32 x 5/8 L.
55-1979	Bag Holder Ass'y. Complete
55-2055	Switch Cover, Bottom
55-2059	Handle, Lower
55-2060	Handle, Upper
55-2066	Blade, 20 Inch
55-2071	Handle Bolt 1/4-20 x 1-1/8
55-2072	Hex Nut 1/4-20
55-2076	Manual
55-2077	Switch Cover, Top
55-2078	Hex Screw
55-2098	Bag, Grass' Catcher
A55-2186	Handle Lock
A55-2187	Handle Bracket Ass'y R. Side
A55-2188	Handle Bracket Ass'y L. Side
A 55-2190	Rubber Washer
55-2194	Cord Clamp
60-1591	Hex Screw 10-32 x 3/4 L.
71-123	Nylon Washer
73-244	Motor Bracket

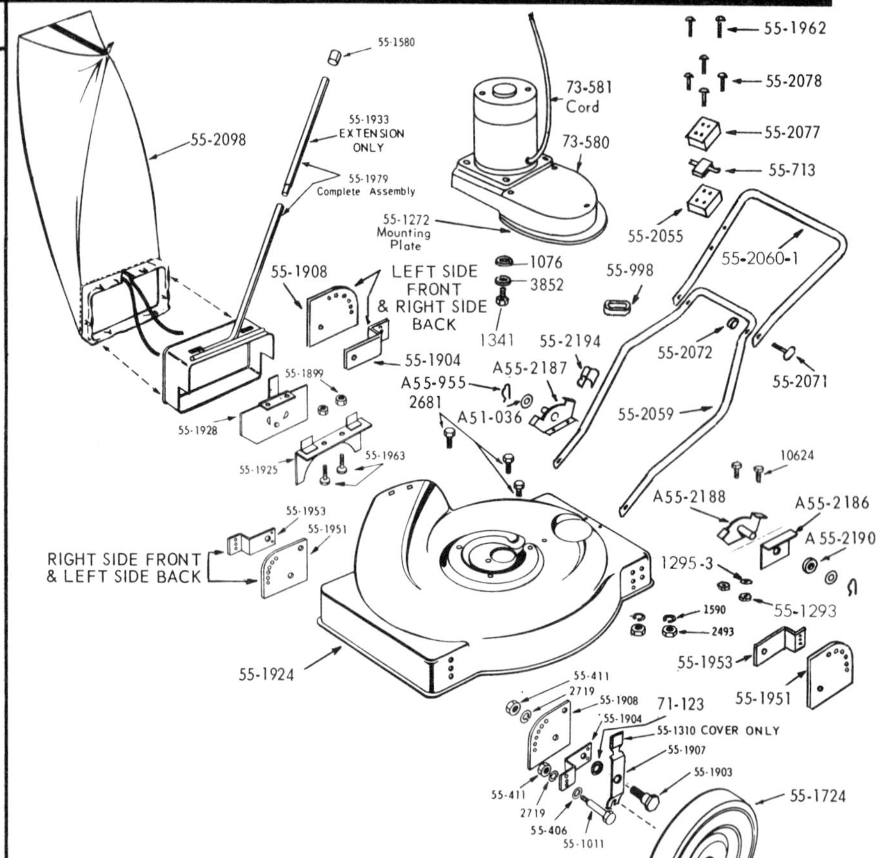

Part No.	Name of Part
73-246	Pulley
73-250	Shaft
73-252	Motor & Belt Drive
73-253	Motor Housing Ass'y.
73-254	Field Coil & Core Ass'y. (W1302)
73-255	Armature Ass'y.
73-257	Baffle
73-258	Motor Support Ass'y.
73-259	Field Coil Ass'y.
73-260	Nameplate (Cover)
73-261	Felt Washer
73-263	Belt
73-264	Pulley Assembly
73-266	Sleeve
73-267	Roll Pin
73-502	Motor Housing
73-504	Field Core Ass'y.
73-513	Washer
73-515	Brush Holder
73-526	Fan
73-532	Drive Shaft Bearing
73-533	Nylon Washer
73-539	Spacer
73-540	Ball Bearing
73-542	Brush Holder Cap
73-543	Motor Cover
73-547	Connector
73-549	Faston Receptacle
73-580	Motor & Belt Housing, Complete
73-581	Cord

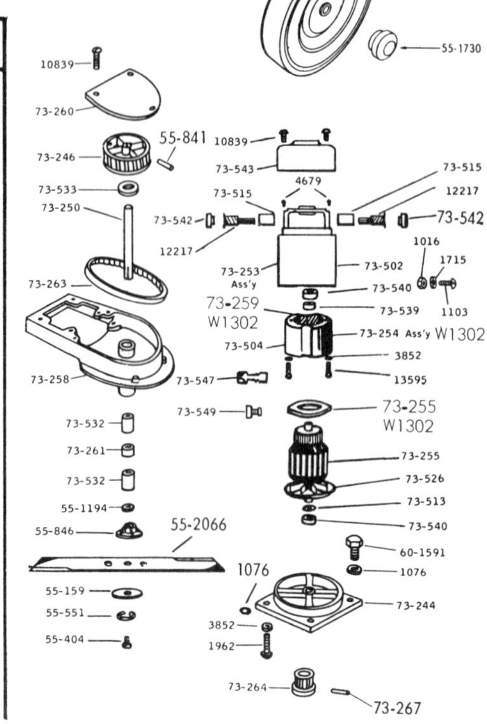

Black & Decker Electric Mower.

Black & Decker manufacture units with both single, and twin-blades. A slight difference in rpm's and horsepower separate the models 8020, 18" twin-blade, 120V AC, 1 1/2 hp, 7500 rpm, 10 amps, and model 8040, 22" 120V AC, 1 5/8 hp, 5200rpm, 10 amps.

The 8040 has cutting height adjustment of 3/4" to 3 1/8", compared to 1 1/4" to 3" on the model 8020.

Both feature easy control adjustment to change cutting height.

All B & D units carry the UL seal of approval, in addition to the PTI - Safety Seal by the Power Tool Institute.

Black & Decker has an extensive dealer organization with service readily available in every section of the United States.

Motor Care.

1. Motor service should be referred to your B & D Dealer.
2. Check cords and plugs to determine where problem might be.
3. Refer to extension cord chart(previous page) for proper cord.
4. Improper cord will cause loss of power and possible overheating.

Blade Care.

1. Previous instructions on blade care apply.
   a. B & D have a blade sharpening grinder wheel that fits in an electric drill. Figure 226.
   b. Grind the blade cutting edge equally to avoid off-balance part , which can cause severe damage to the motor.
   c. B & D replacement blades are quite inexpensive.

Figure 226

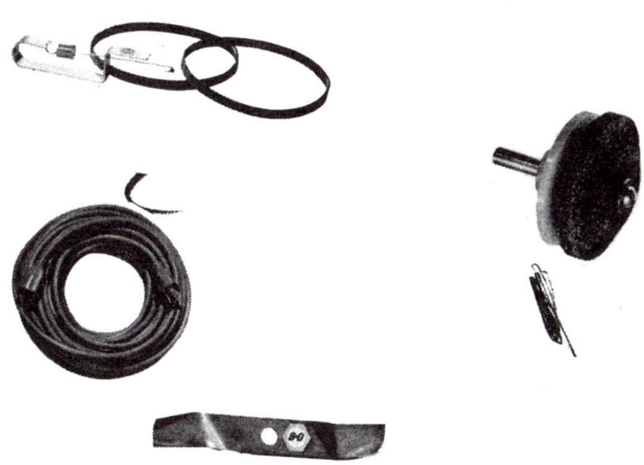

Black & Deckers "Cordless 18"

B & D have marketed a cordless electric lawn mower which is midway between the cordless razor, and the electric cart-fork lift area in concept and operation.

The power plant and replacement battery are a beautiful combination. I have been in the materials handling field for years and am fully acquainted with battery powered warehouse equipment. B & D have reduced the size, but retained the necessary power to cut 10, 000 square feet of lawn on single charge, under normal operating conditions.

Model 8080 Cordless 18.

1. The switch is momentary-contact type. Press to go, release-stop.
2. To change cutting height, simply move control(c) to setting.
3. Cutter bar adjustment can be made with micrometer accuracy(d).
4. The unit has rear wheels in lieu of rollers, and they're calibrated, for simple adjustment.
5. To sharpen the reel-blade/bed knife, simply change the plug position and the reel turns in reverse.
   a. As in previous instructions, set bed knife closer to reel than in normal operations and run mower for several minutes.
   b. Apply the oil and grinding compound mixture if blades are very dull. Wipe away all excess after adjustment.
6. The mower motor can be recharged on unit furnished with the 8080.
   a. Plug charger into standard 120V.

Figure 227

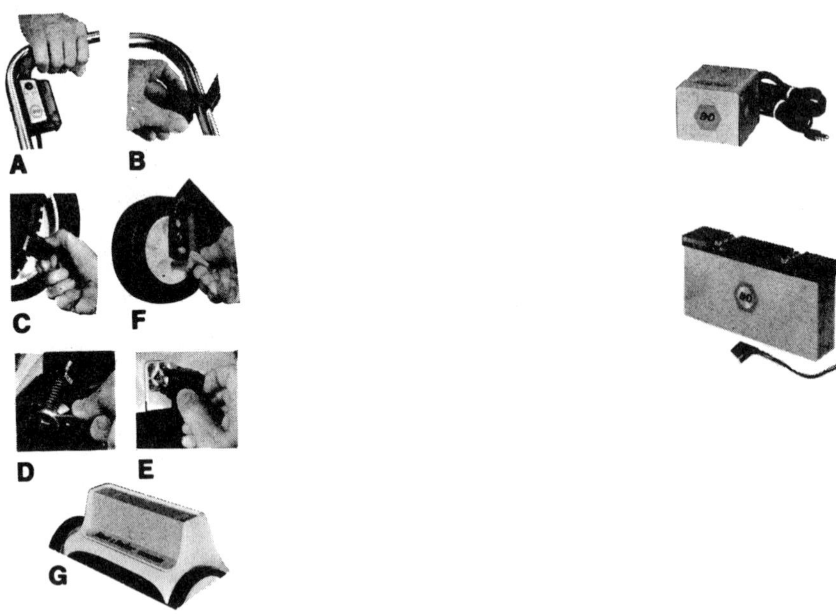

Chapter Eleven

Safety

Statistics covering home accidents indicate the lawn mower is a leading contributor year after year. Most involve power mowers and the foremost reason is just plain carelessness.

Most manufacturers covered in this book, and all leaders in the lawn - care field, conform to safety standards of organizations such as O. P. E. I. (Outdoor Power Equipment Institute, Inc.), or ASA, safety association.

The following are safety suggestions for power mower users.

Training.

1. Regard your mower as a piece of power equipment.
2. Never allow children or young teenagers to operate power mower.
3. Be sure you know how to stop mower & engine at moments notice.
4. Keep children away from operation area at all times.

Preparation.

1. Clean entire lawn of all debris before mowing.
2. Plan cutting pattern to avoid pulling mower back toward you.
   a. If mower is rear wheel drive, use caution when changing angle or backing downward, and in most cases, disengage clutch.
3. When mowing in rough terrain, or high weeds and grass, set blade at highest cutting position to minimize debris ejection from mower.
4. Restrict rotary mowers in particular to areas free of debris.
5. Mow during daylight hours unless the area is well lighted.
6. Do not operate power mower in wet grass. Wet grass clogs mower and increases operating danger to user.
7. Fill gasoline-driven mower outdoors. Avoid spilling gasoline, and don't fill tank with engine running, or while you're smoking.
8. Don't mow while barefoot or in open sandals. Wear heavy shoes.
9. Don't remove mower shields while in operation.

Operation.

1. Pay attention. Don't become distracted.
2. Keep area clear of all persons, especially small children.
3. Don't start engine and blade until you're ready to mow.
   a. Stop the engine whenever you leave the mower.
4. Don't overspeed engine. Operate with engine speed at slowest point to do an efficient job. Don't tamper with governor setting.
5. Start mower carefully. Plant feet firmly, away from blade, before activating the starter. Be sure mower won't tip or roll during start.
6. If mower handle contains safety latch, keep it latched at normal operating position during operation.
7. Keep in step with mower. Don't lag behind or let it pull you along. Maintain full control at all times. Don't run while pushing mower.
8. Control mowing direction by hand pressure on handle, not with foot pressure on the mower housing.
9. When operating the mower, use the handle to push or pull the mower and not on the deck with your hands or feet.
10. When operating over uneven terrain and slopes, use extreme care, and be sure of solid and firm footing at all times.
11. Avoid mowing over objects, use care and avoid objects with blade.

Safety, continued.

12. Stop operation when another person approaches. Don't stand on the grass discharge side of mower with engine running.

Maintenance and Storage.

1. Follow manufacturers recommendations for maintenance.
2. Have mower serviced at least once a year.
3. Never adjust mower or change attachments unless engine is off, and spark plug wire is disconnected. Never work near blade or cutter bar, unless this precaution has been taken.
4. If carburetor adjustment is necessary, stand clear during service.
5. Keep engine free from foreign matter such as grass, or grease and oil drippings. An accumulation of these items may result in a fire.
6. Store gasoline in a safe container, and in a cool, dry place.
7. Keep mower and fuel container in locked storage to keep from children or pets.
8. Maximum mowing results and safety can only be expected if mower is maintained and operated correctly.
9. Proper storage procedure will prevent gum deposits which are very harmful to the engine.

Storage. The following steps will apply to a majority of mower engines.

1. Start engine and close shutoff valve, which drains gasoline from the carburetor.
2. Disconnect fuel line from shutoff valve and remove shutoff valve from tank. Turn off counterclockwise. Excess gasoline will drain out.
3. Clean shutoff valve screen, and replace shutoff valve with lever to open position.
4. Change oil while engine is warm. Follow previous instructions.
5. Remove the spark plug and pour 1/2 ounce of oil into cylinder. Turn engine slowly, by hand, to allow oil to spread over entire area.

ACKNOWLEDGEMENTS

Sincere Press is extremely grateful for assistance, materials or drawings supplied by the following persons or firms.

Jacobsen Manufacturing Company, 1721 Packard, Racine, Wisconsin.
Southland Mower Company, Box 347, Selma, Alabama.
Pioneer Gen-E-Motor Corp. 5841-49 Dickens Av. Chicago, Ill.
Sensation Mfg. Co. 7577 Burlington St. Ralston, Nebraska.
Ariens Company, 109 Calumet, Brillion, Wisconsin.
Montgomery Wards, Chicago, Ill.
AMF Western Tool Div. Des Moines, Iowa.
Clinton Engine Company, Maquoketa, Iowa.
Steen's Inc. 1635 W. Valley Blvd. Alhambra, Calif.
Tecumseh Products Co. Grafton, Wisc.
Neal's Lawn Mower Service, Phoenix, Az. (Mr. Neal Leiter).
Western Auto Stores, Tucson, Az. Gilbert Mowbray, Mgr.
Southwest Auto Supply, Tucson, Az. Gene Fest, owner
Erv's Equipment Rental, Tucson, Az.
Robert Crawford, Tucson, Az.